# Theory and Evaluation of Formation Pressures

*The EXLOG Series of Petroleum Geology and Engineering Handbooks*

# Theory and Evaluation of Formation Pressures

## *A Pressure Detection Reference Handbook*

**Written and compiled by EXLOG staff**

**Edited by Alun Whittaker**

International Human Resources Development Corporation • Boston

**Library of Congress Cataloging in Publication Data**

Main entry under title:

Theory and evaluation of formation pressures.

(The EXLOG series of petroleum, geology, and engineering handbooks)
Bibliography: p.
Includes index.
1. Reservoir oil pressure. 2. Petroleum—Geology.
I. Whittaker, Alun. II. EXLOG (Firm) III. Series.
TN871.T44 1985 622′.1828 85–2287
ISBN 0–88746–052–6
[90–277–1979–9 D. Reidel]

Printed in the United States of America

# CONTENTS

# ILLUSTRATIONS

# PREFACE

The objectives of this book are: (1) to educate the prospective Pressure Evaluation Geologist to a basic level of expertise; (2) to provide a reference tool for the experienced geologist; and (3) to foster constructuve thought and continued development of the field geologist.

Despite the incorporation of many new ideas and concepts, elaboration of the more recent concepts is limited due to space considerations. It is hoped that the geologist will follow up via the literature referenced at the end of each chapter.

Easy reference is provided by the detailed table of contents and index. A glossary of terms, definitions, and formulae adds to the usefulness of this reference text.

# ACKNOWLEDGMENTS FOR FIGURES

Figure 2–7 is reprinted by permission of the AAPG from Barker, 1972.

Figure 4–21 is courtesy of Totco

Figure 5–2 is reprinted by permission of the *Oil and Gas Journal* from Matthews and Kelly, 1967.

Figure 5–4 is reprinted by permission of the SPE-AIME from the *Journal of Petroleum Technology* from Eaton, © 1969.

Figure 5–5 is reprinted by permission of the SPE-AIME from Hubbert and Willis, © 1957.

Figures B-1—B-5 are courtesy of Schlumberger Well Services

Figure C-3 is reprinted by permission of the SPE-AIME from the *Journal of Petroleum Technology* from Jordan and Shirley, © 1966.

# 1 INTRODUCTION

## 1.1 WHAT IS PRESSURE EVALUATION?

Pressure evaluation is an integral part of formation evaluation. The physical, chemical and geological processes which contribute to the formation of pore pressures require varying time and space, and the result is that pore pressures range from overburden magnitudes to as low as 30 to 40 percent of normal hydrostatic pressures.

By using the modern and accepted concepts outlined in this guide, relationships between geology and the engineering processes may be interpreted to give an accurate estimation of the pore pressure at any point in the drilled formations. The application of the methods, procedures and knowledge required to arrive at a complete record of pore pressure trends and values is the major component of geopressure evaluation. Other factors depend upon the services available in the logging unit, the requirements of the client, and the conditions encountered as the well is drilled. These responsibilities may involve electric log interpretation, pre-drill analysis of areal and offset well data, engineering suggestions while drilling, communication of reports on a daily and weekly basis, and the compilation of a Final Well Report for presentation to the client.

## 1.2 FROM FORMATION LOGGING TO PRESSURE EVALUATION

A Pressure Evaluation Geologist (P.E.G.) or GEMDAS® Operator is first an experienced logging geologist. He has gained a thorough understanding of rig procedures, personnel relationships, basic and advanced mud logging techniques, and interpretation. He has possibly been exposed to pressure evaluation methods either by logging in GEMDAS units or by having performed basic logging services in areas of geopressure occurrence.

The logging geologist's basic understanding and experience must then be utilized to the fullest extent. His further education is attained through continuing experience, studying this guide, and participating in special training programs in order to achieve the level of expertise expected of the Pressure Evaluation Geologist. As a P.E.G., he will use the data collected and recorded by the logging geologist. The two activities are by necessity interrelated: the Pressure Evaluation Geologist cannot fulfill his tasks without lithological and gas data, and the logging geologist must take into account the effect of the pore pressure/mud pressure relationship in his interpretation of gas shows and cuttings analysis.

The GEMDAS operator has access to a greater range and quantity of data, and has powerful computer hardware and software to process them. Combining this with geological observations and his P.E.G. experience, a refinement in the pressure evaluation service is possible.

GEMDAS® is an Exlog registered service mark, standing for Geological and Engineering Monitoring and Data Analysis Service.

## 1.3 RESPONSIBILITIES

When a logging geologist is promoted to a Pressure Evaluation Geologist, he accepts a great responsibility because the decisions and reports made in the course of his duties are of importance to the drilling operations as a whole. Reports should thus be accurate, subject to critical examination in difficult situations, and, most important of all, they must be able to be substantiated.

Both the P.E.G. and GEMDAS operator work in very close cooperation with the Operator's engineer and geologist, the rig superintendent and toolpushers, the mud engineer, and the Operator's representatives at the local base or in town. The ability to communicate with these personnel is vital.

During the performance of his duties the Pressure Evaluation Geologist will find that some wells are trouble-free and very undemanding; however, this is no reason to reduce the quantity or quality of his observations and records. Conversely, some wells or intervals will place enormous stress and responsibility upon the geologist such that his knowledge and capabilities will be tested to the utmost. Every well is different, and knowledge may be gained from every wellsite situation. The completion of a demanding assignment which results in the attainment of total depth with the minimum amount of hole problems and the maximum amount of information is one of the most rewarding aspects of the job.

## 1.4 INSTRUMENTATION

The P.E.G. will be trained to use various calculators and computers which utilize cassette, disk or tape drives and printer/plotters. They are an invaluable aid in making pressure calculations, and several Engineering Assistance Programs (EAPs) are available to the P.E.G.

Instrumentation used to assist in geopressure evaluation will vary depending on the particular job. The minimum "pressure package" is a standard logging unit equipped with an HP9820 and magnetic cards. More sophisticated configurations range from a logging unit with a DMP, MFE recorder with HP9820, cassette drives and a printer, to the fully equipped GEMDAS units with computers, tape drives or disks and printers (see Figure 1-1).

Generally, the less equipment, the more running about the geologist has to do, but this does not mean that in the fully equipped units the geologist has to confine himself to the unit. Also, the final results and amount of relevant data collected should be the same, regardless of the level of complexity of equipment.

Secondary equipment, such as Mudweight-In and -Out, Mud Flow Comparator, Temperature-In and -Out, CO2, H2S, All Gas Detector, PVT, Shale Density, Shale Factor, Calcimetry and Conductivities, may or may not be available depending on the contract with the client. But the ability to use, interpret, calibrate and troubleshoot all unit equipment is necessary. Refer to the appropriate hardware and applications manuals for information.

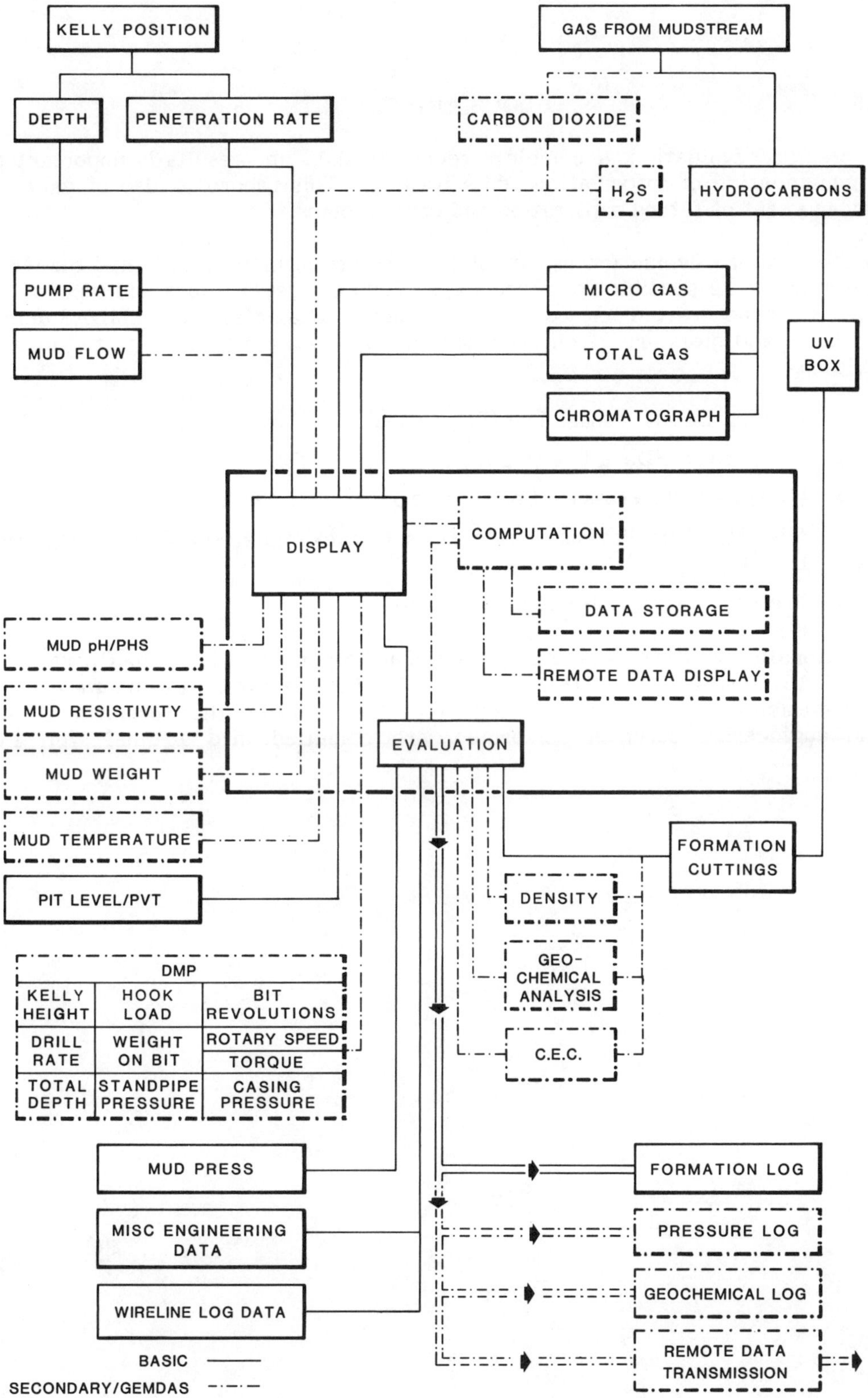


Figure 1-1. Logging unit systems and information flow

## 1.5 LOGS AND REPORTS

In pressure evaluation, a complete record of data and results is important to the communication of information while drilling. This record is also of value in the development of future exploration and drilling plans.

In the pressure evaluation aspect of their work, both the P.E.G. and the GEMDAS operator are responsible for preparing a group of pressure logs and reports. These may be hand-drafted or computer-plotted. Examples of GEMDAS computer printouts and plots are given elsewhere in the text. The hand-drafted logs are as follows:

- Drilling Data Pressure Log (Figure 1-2)
- Temperature Data Log (Figure 1-3)
- Wireline Data Pressure Log (Figure 1-4)
- Miscellaneous Pressure Data Logs (for example, the Shale Data Pressure Log, Figure 1-5)
- Pressure Evaluation Log (Figure 1-6)

Additionally, pressure data and comments are reported daily to the Operator on the GEMDAS Logging Report (Figure 1-7). At the completion of the well all of the information in these reports and logs, together with the other formation evaluation and engineering services performed, are compiled into a Final Well Report.

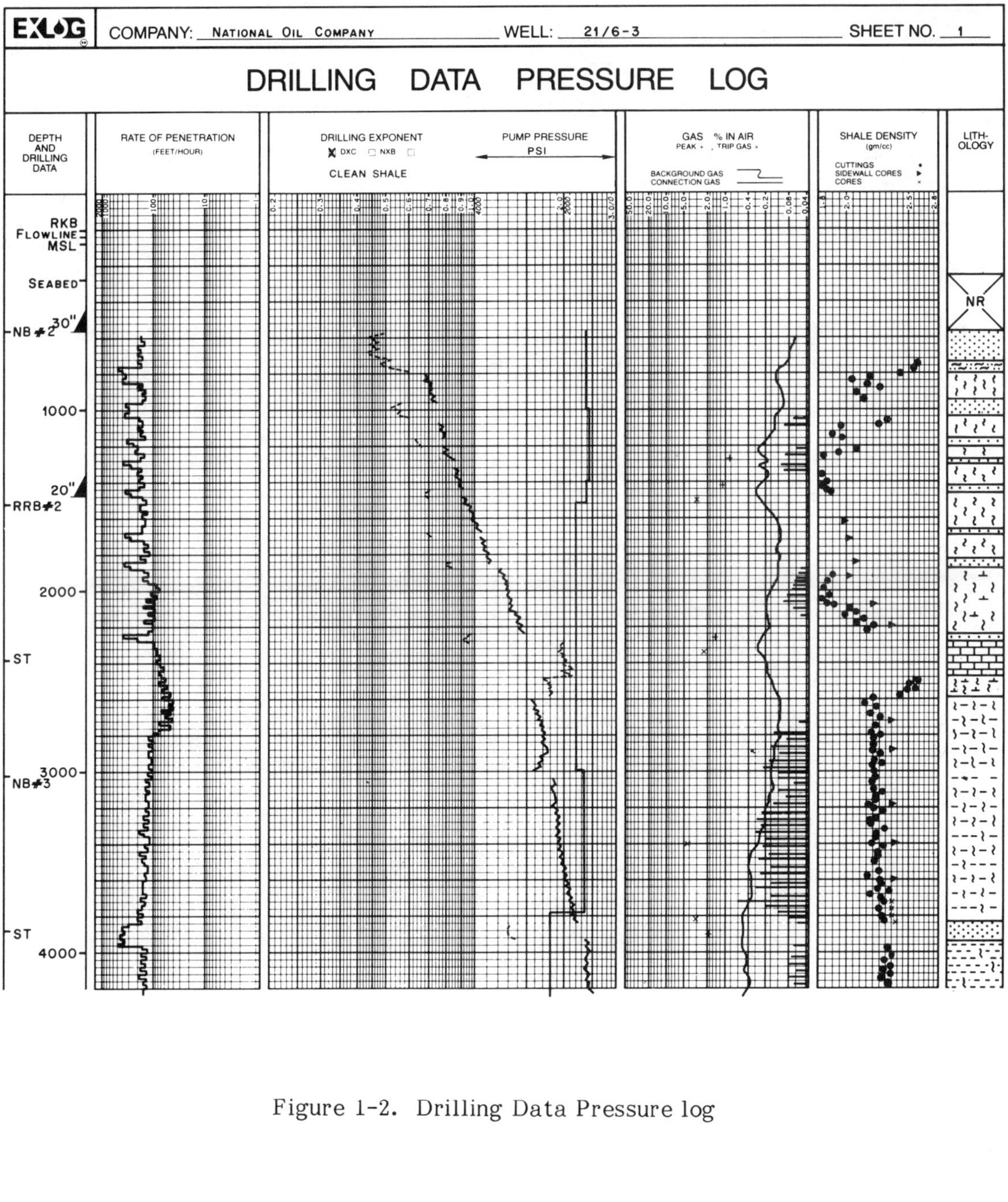


Figure 1-2. Drilling Data Pressure log

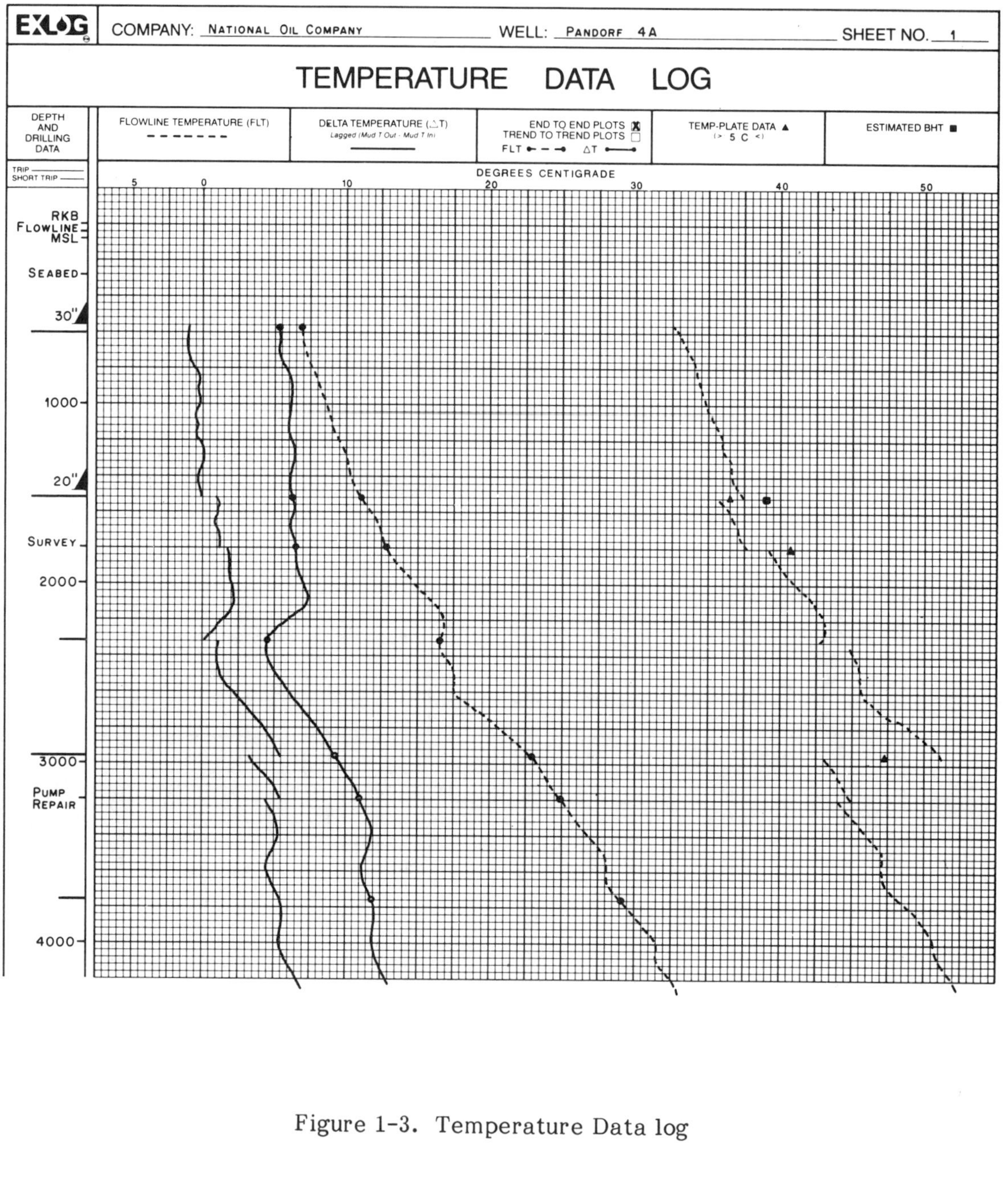


Figure 1-3. Temperature Data log

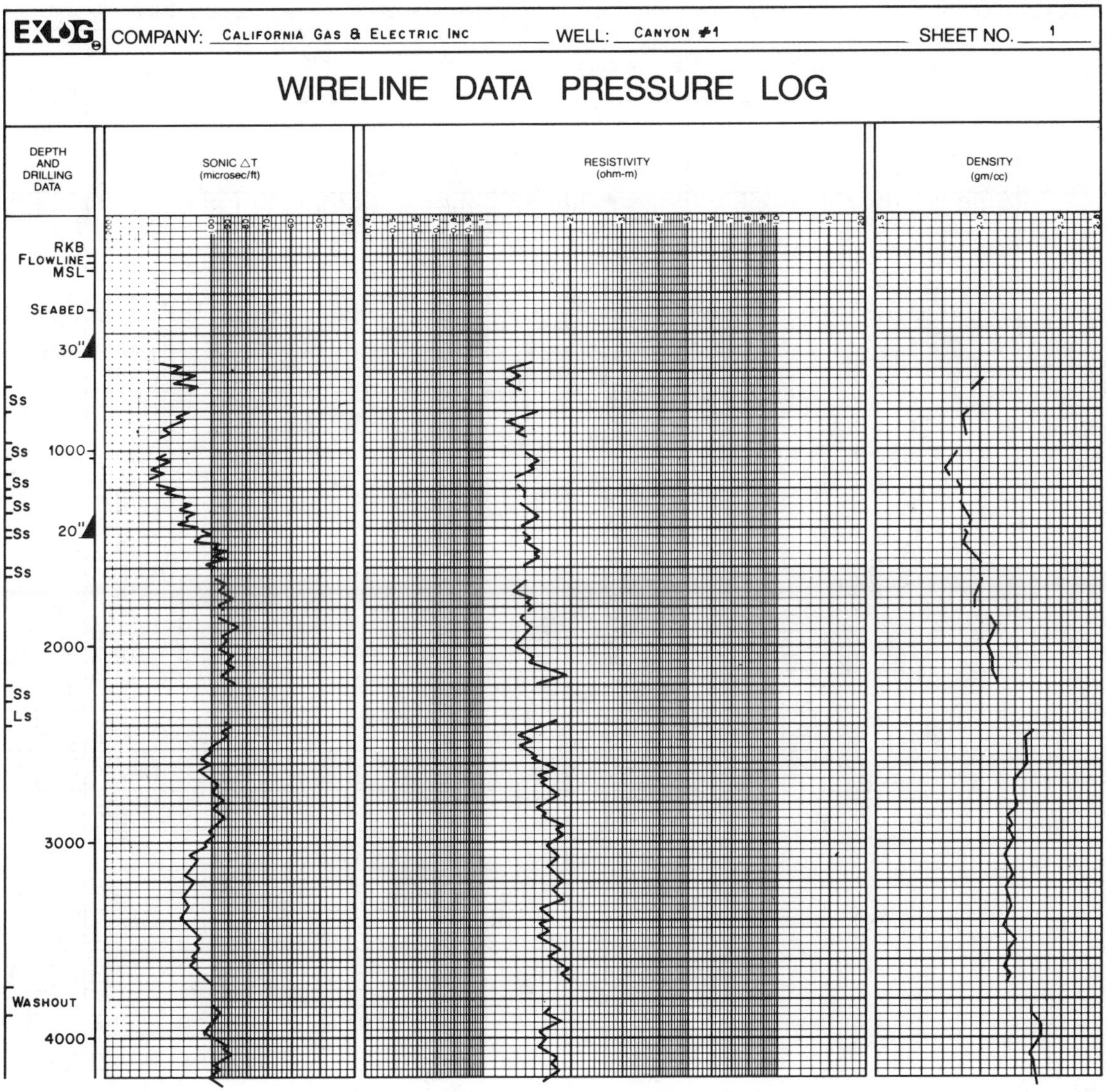


Figure 1-4. Wireline Data Pressure log

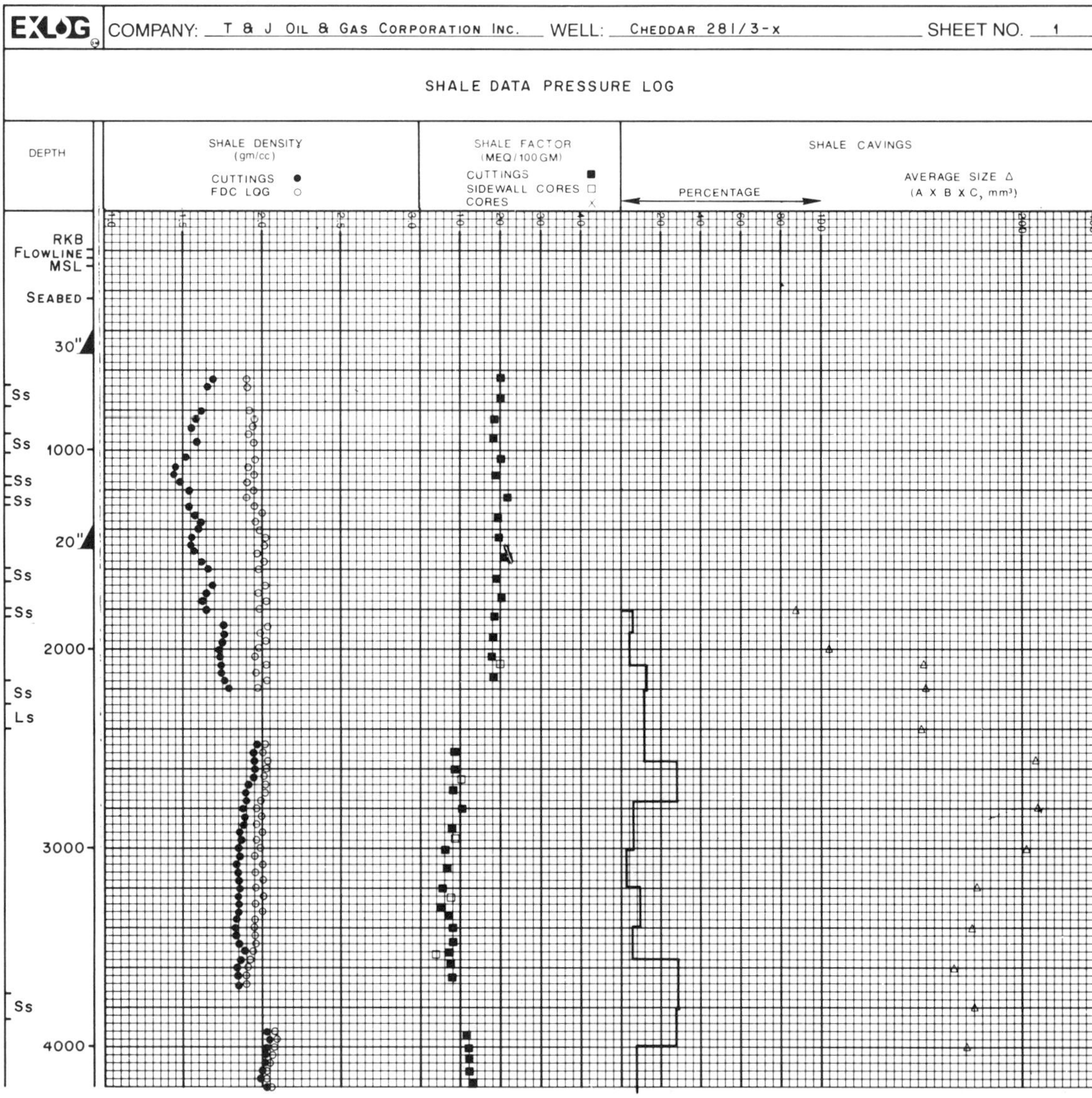


Figure 1-5. Shale Data Pressure Log

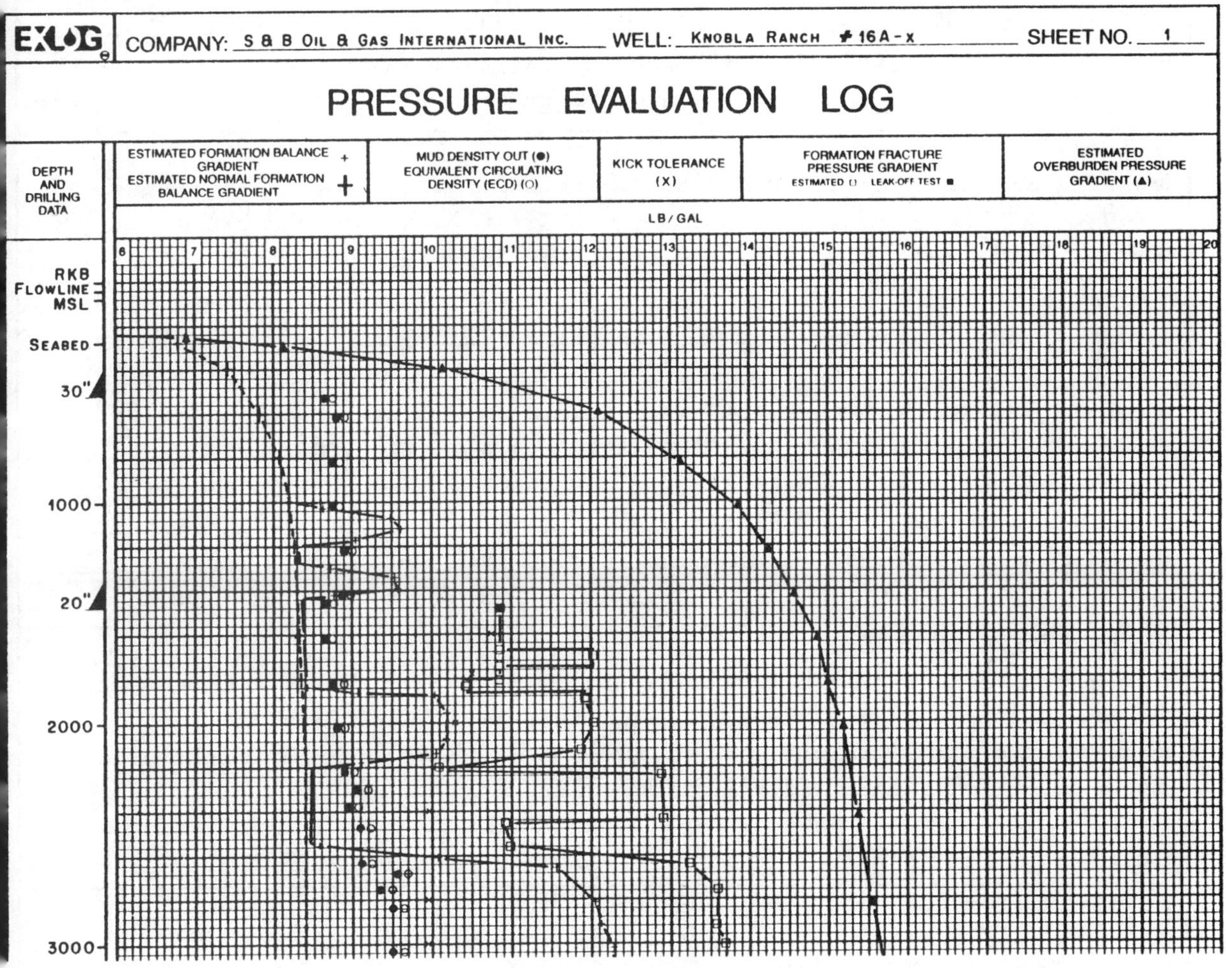


Figure 1-6. Pressure Evaluation log

EXLOG (SM)

# GEMDAS LOGGING REPORT NO. 27

COMPANY Plough Resources Inc. WELL Shield 211/12-7

DATE 1/27/80 TIME 0600, Morning Report

DEPTH 11765 ft LAST REPORT DEPTH 11665 ft

RIG OPERATIONS Drilled w/ NB#17 thru blocky gray shales, slightly calcitic.

REPORT BY Watkins REPORT RECEIVED BY ______ (OPERATOR) SIGNED

## DRILLING REPORT

Bit No.: 17 Type: Reed FP51 Size: 8½" Jets: 11x11x12

On Bit: Footage: 462 Hours: 17½ ROP: 6–27 fph WOB: 40 Klb RPM: 120

Pump Press: 2660 SPM: 2x60 Torque: 4 Kft-lb TBR: 107921 CPFT: $ 13–35 CPFB: $ 83

## HYDRAULICS REPORT

Mud Density In: 10.1 Mud Density Out: 10.0 lb/gal ECD: 10.3 PV/YP: 16/12

Gels: 5/7 Salinity: 67000 PPM Cl Solids: 12 %

Hole Volume: 993 bbl Annular Volume: 648 bbl Tubing Volume: 226 bbl Displaced Volume: 119 bbl

Carbide Lag—Calculated Lag: 5768–5547=221 str=41 bbl Flowrate: 850 gpm

Drillpipe Annular Vel (Max. Dia. Sec.): 24 ft/min Drillpipe Annular Vel (Open Hole): 153 ft/min

Drill Collar Annular Vel (Open Hole): 356 ft/min Critical Vel: 368 ft/min

Pressure Loss System: 280 psi Pressure Loss Bit: 2380 psi % Pressure Loss: 87%

Nozzel Vel: 531 Jet Impact Force: 788 HHP: 404

## PRESSURE PARAMETERS

Drilling Exponent: (Nx): Trend=1.17; Avg.=1.17 Flowline Temperature: FLT=46.3 C°; ΔT=2.6 C°

Shale Density: Trend=1.75; Avg.=1.96 Shale Factor: Trend=25; Avg.=26

Background Gas: .32–.46% Max. Formation Gas: .86% @ 11735 ft Trip Gas: ---- @

Other Gas: Regular connection gases from 11690 ft, .05–.11%

Fill: None Tight Hole: None

Cavings: Est %: Less than 10% Average Size: 7mm x 5mm x 3mm

## ESTIMATED PORE AND FRACTURE PRESSURE

Kick Tolerance: 12.6 lb/gal Min. Estimated Fracture Pressure (Open Hole): 12.9 lb/gal

Estimated Pore Pressure: 8.6 lb/gal Min. Estimated Pore Pressure (Open Hole): 8.6 @ T.D.

Max. Estimated Pore Pressure (Open Hole): 8.9 @ 11380 ft Estimated Fracture Pressure at TD: 12.9

Comments:

All pressure parameters suggest that pore pressure remains close to normal. High shale densities due calcite content.

Increased torque from 11760 ft, suggests formation may becoming sandy. Sample not yet at surface.

EL P/N 18429 MAY 1980

THIS REPORT IS GOVERNED BY THE TERMS AND CONDITIONS AS SET FORTH ON THE REVERSE SIDE

Figure 1-7. GEMDAS Logging Report form

# 2
# GEOLOGY

## 2.1 INTRODUCTION

Pore pressures can be "normal," i.e., simply the pressure exerted by a column of water, or they may be "abnormal" or "subnormal." Normal pore pressure at a point in the geological section will be the hydrostatic pressure due to the average density, $\overline{WF}$, and vertical depth, Dv, of the column of fluids above that point — that is, to the water table, onshore, or sea level, offshore (see Figure 2-1). The convention is that abnormal pore pressures are higher than normal, and subnormal pore pressures are lower. The term "geopressure" applies specifically to fluid pressures that are in excess of the normal hydrostatic pressure. The mechanisms by which abnormal pore pressures are formed have been debated for a number of years.

In order to explain the formation of some geologic structures, geologists surmised that some form of excessive pore pressure must have existed in a specific horizon at the time of deformation. Structures like the Tay Nappe, Moine Thrust, overthrust belt of Western Wyoming, Idaho and Utah, Naukluft Mountains in Namibia, and the Upper Helvetic Nappes of the Alps must have been able to move over a surface of extremely low friction so that coherence was maintained. It is

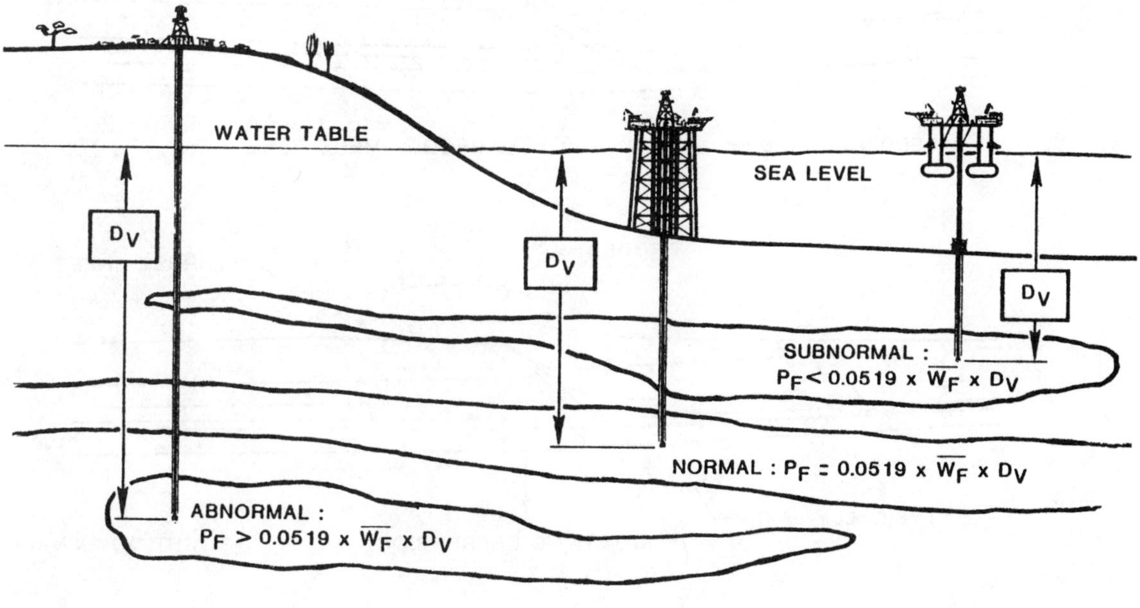


Figure 2-1. Pore Pressure — Normal, Abnormal and Subnormal

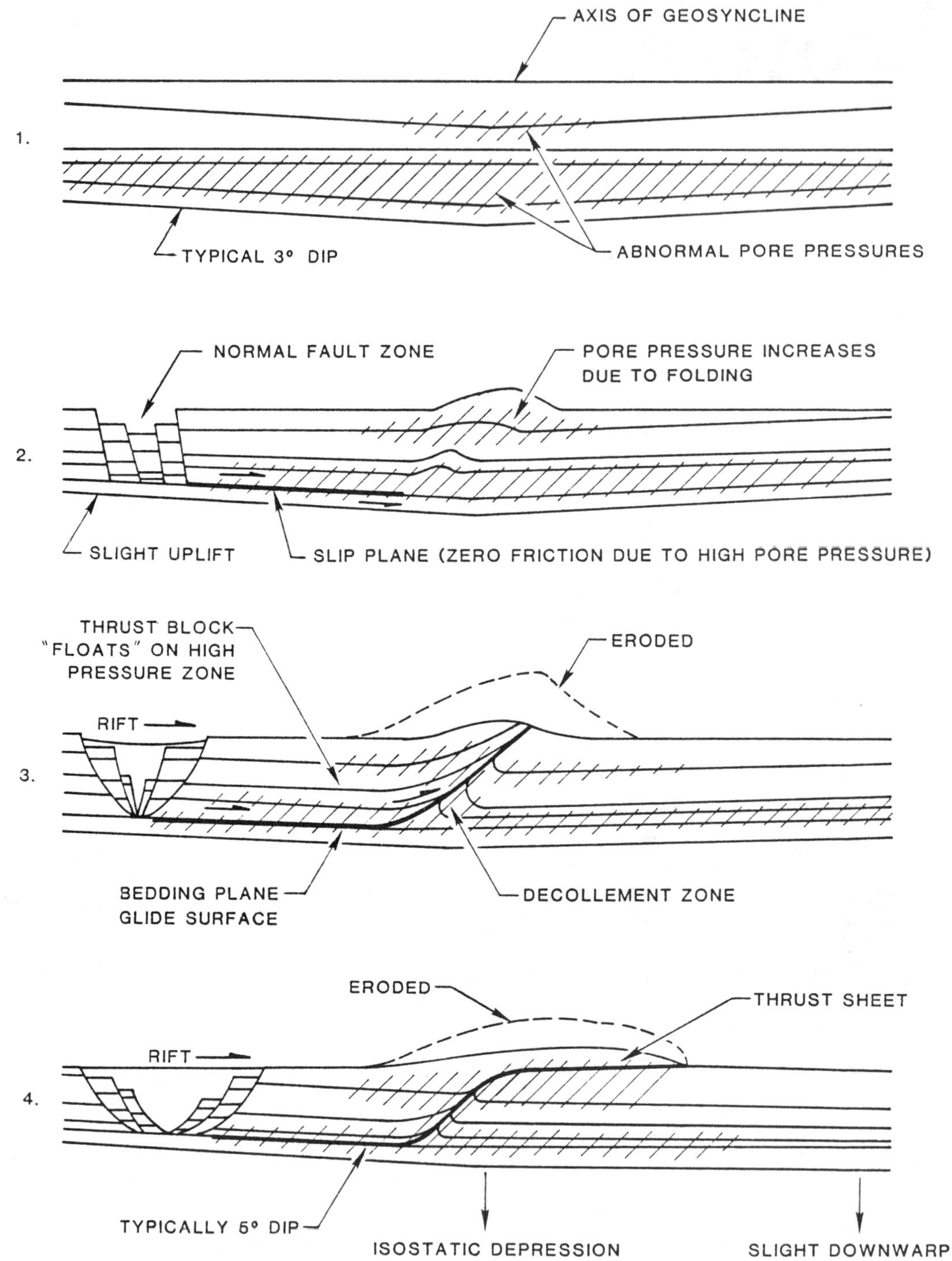


Figure 2-2. Idealized diagram of zones of abnormal pore pressures and development of overthrust and thrust sheet on the flank of a geosyncline

believed that the "zero friction" necessary was due to the effect of the pore pressure being so high that the overlying strata were able to "float" on a hydraulic cushion. This high pore pressure reduced the shearing strength of the rock to a negligible quantity such that, upon application of a lateral tectonic force, the strata overlying the high pressure zone were free to move. On the flanks of geosynclines, due to basinal subsidence, the sediments may dip at a small angle, approximately 1 to 3 degrees. In some cases (for example, the Wyoming overthrust belt), it is thought that the force of gravity would have been sufficient for movement of the strata overlying the high pressure clays, down dip.

Typical displacement of these thrust zones are of the order of several kilometers, and the displaced sedimentary thicknesses vary from 10,000 to 25,000 feet. Thin section examination of the rocks from the base of the thrust generally shows very thin brecciation, if present at all. In most cases, the bases of the various thrusts are remarkably undeformed: there is little evidence of strained and shattered grains, and the rocks are unmetamorphosed. Some graphite may be present in carbonaceous shales, and some secondary mica has been observed in shale, indicating that only minor heat-induced recrystallization has occurred. In view of the tremendous distances involved in moving these allochthons, abnormally high pore pressures would provide a means to account for the observed anomalous phenomena. Figure 2-2 shows a typical thrust zone development sequence.

Although a zone of abnormally high pore pressure was postulated to have existed and to have been implemented in the genesis of these structures, the mechanism and origin of the geopressures were unknown. Visible evidence on a smaller scale in the form of veins filled with calcite or quartz provides a record of palaeo stress orientations which resulted in the "permanent" finite strain structures. Again, very high pore pressures were assumed to have been instrumental in forming these specific types of veins, through the action of a tectonic trigger — autohydraulic fracturing. Figure 2-3 illustrates these vein types.

The above discusses the importance of very high pore pressures with respect to geological processes, but very low to nonexistent pore pressures also play an important role in the generation of geological structures. Fault movement and geometry appear to be closely related to the pore pressure in the adjacent lithons.

The type of fault movement has been shown to be dependent on the magnitude of the pore pressure in the adjacent lithons. This is of major importance since a continually active seismic fault has a considerable effect on the habitants and economics of the zone, whereas an aseismic fault causes no such major problem.

A fault of long activity is defined as one that has been seismically active for the past 60 million years (such as the Anatolia fault in Turkey). The continual seismicity is evidence that mylonitization may be occurring, and if movement is rapid, pseudotachylite generation may occur on the fault surfaces where there is zero pore pressure. Recognition of these structures in the outcrop then gives supportive evidence that the fault zone was seismic, and, if pseudotachylite is present, the pore pressure in the fault plane and adjacent lithons was necessarily zero, i.e., zero porosity.

Significant parts of the San Andreas fault are aseismic. Steady, consistent movement of 2 to 4 centimeters per year occurs without any stick-slip type displacements. The mechanism for this steady movement is still not fully understood; however, the presence of serpentinites and high pore fluid pressure in the area around the aseismic zones are possible agents that assist aseismic movement.

Pore pressures thus play a major role in the formation of major structures in sedimentary rocks. The magnitude of the pore pressure in a given stress field determines the form of the resulting structure. Structure produces topography and hydrocarbon traps, and it influences sedimentary environments. Since pore pressures are related so closely to deformation processes, and since through oil exploration it is possible to encounter, measure and examine the various pore pressure/lithological relationships, a thorough understanding of the generation of pore pressures in sedimentary rocks provides a major step toward increased expertise in the interpretation of geological processes.

The following paragraphs outline current theories on the generation of geopressures. Petroleum exploration has played a major part in the development of these ideas, and thus its importance in the comprehension of geological processes is significant.

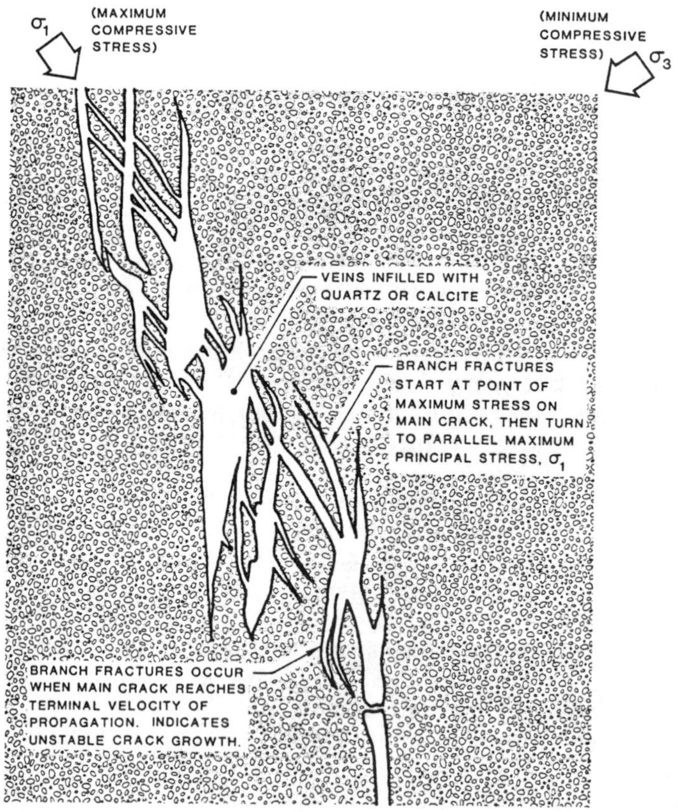


Figure 2-3. Typical en-echelon hydraulic fractures found in stressed or tectonically deformed rocks

## 2.2 COMPACTION DISEQUILIBRIUM

When the upper layer of freshly-deposited argillaceous sediment is undergoing initial compaction at and just below the sediment/water surface, the interstitial fluid is continuous with the overlying seawater and hence the pore pressure is essentially hydrostatic. If this sediment is composed of hydrated clays, then due to the presence of adsorbed water layers the solids will not have direct physical contact. As the upper argillaceous sediment layer becomes buried under further layers of sediment, a gradual compaction takes place. If the sedimentation rate is slow, this layer will gradually adjust to the additional load imposed by the overlying sediments. As the mineral grains are pressed closer together, pore water is expelled. With time, pore pressures will also remain in equilibrium if there is adequate fluid intercommunication between grains and adjacent sediment layers. Argillaceous sediment has high porosity and is permeable in this initial state, thus the expelled fluids will flow ultimately in the direction of least resistance, i.e. upwards or to a lower pressure potential in a porous sand layer. As long as the fluid can escape under normal sedimentary conditions, and the permeability is sufficient, hydrostatic pore pressures will prevail.

For this equilibrium to be maintained, a delicate balance is necessary among (1) the rate of sedimentation and burial, (2) the magnitude of the permeability, (3) the rate of reduction of pore space, and (4) the ability of the excess pore fluid to be removed. If the rate of sedimentation is very slow in comparison to the other processes, hydrostatic pressures are maintained. Also, if the availability of porous/permeable conduits is high (for example, in clay/sand interbeds), then hydraulic flowpaths are efficient and normal dewatering can occur, provided the sand aquifers possess communication with lower fluid potentials.

The initial porosity of clays is generally between 60 and 90 percent, depending on the major clay type. Hydrated, flocculated clays have the highest porosity. Plastic, wet clays may have a porosity of 40 to 50 percent, and compacted clay/shale will have a porosity of less than 15 percent. This tremendous decrease in porosity from the point of deposition to complete lithification indicates the vast amount of water that must be removed from a thick clay sequence during its burial history. Theoretically, pore water expulsion from a compacting clay sequence follows a curve similar to that shown in Figure 2-4.

If one of the processes which allow equilibrium to be maintained during steady burial is restricted, the expulsion of excess pore water will be impeded. In time, the result is an increase in pore pressure. Given a normal rate of sedimentation of argillaceous sediment, the permeability and rate of compaction remain relatively constant for each arbitrary compaction unit. Should a change in sedimentation rate occur, such that the amount of clay deposited in unit time exceeds the amount that produced equilibrium conditions, the process of dewatering would be restricted because the permeability has remained essentially constant. Thus the magnitude of sedimentation increase over equilibrium conditions is proportional to the degree of pore pressure increase. Also, as the excess pore fluids help support the increasing overburden load, further compaction is retarded; or if expulsion has ceased, further compaction will not occur until hydraulic communication is reestablished. Hence the formation becomes abnormally pressured because the fluids are subjected not only to hydrostatic forces but also to the weight of the newly deposited sediment.

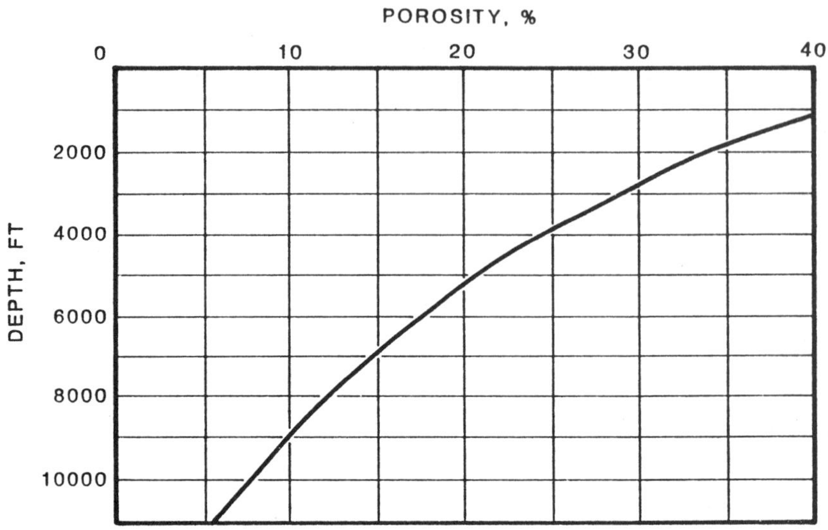


Figure 2-4. Porosity/Depth relationship for a typical compacting clay sequence

Areas where rapid sedimentation of argillaceous sediment occurs are typically continental margins, and more specifically in deltaic regions. In these areas it is likely that the sedimentation rate has exceeded that which would ensure equilibrium, hence it would be common to encounter abnormal pore pressures.

Where a zone of high pore pressure exists within an argillaceous sequence, it exhibits different properties from the normally compacted sediment around the high pressure zone. Due to its high pore pressure, the sediment has a relatively high porosity and hence a relatively low bulk density because of the excess trapped pore fluids. In a normal compacting sequence, bulk density increases (with decreasing water content) to a maximum defined by the densities of the constituent minerals when porosity is zero. In actuality this end-point is never reached: all rocks have some porosity. It may be concluded, therefore, that if a uniform clay sequence is penetrated which exhibits a zone of decreasing bulk density with depth, the low density zone has relatively higher porosity and pore pressure than the normal sequence above. Figure 2-5 shows this density reversal relationship with depth (curve A). The low density can be equated to the depth at which that density would normally be expected, and this is termed the equilibrium depth. This occurs at point 1. Theoretically it may be assumed that at point 1 dewatering ceased completely (mechanism ignored) so that the additional overburden load from further sedimentation was borne entirely by the pore fluid. This theoretical "compaction" path is shown as curve B. A necessary assumption for this particular model is an instantaneous, complete isolation of pore fluids at point 1. A far more likely compaction path is shown as curve C, which indicates (1) that compaction has been slower than the increase in overburden load due to rapid sedimentation, and (2) that the permeability was such that removal of all of the excess pore water could not occur. This compaction path probably represents reality quite closely, in that slight changes in mechanisms are more likely than an abrupt instantaneous change.

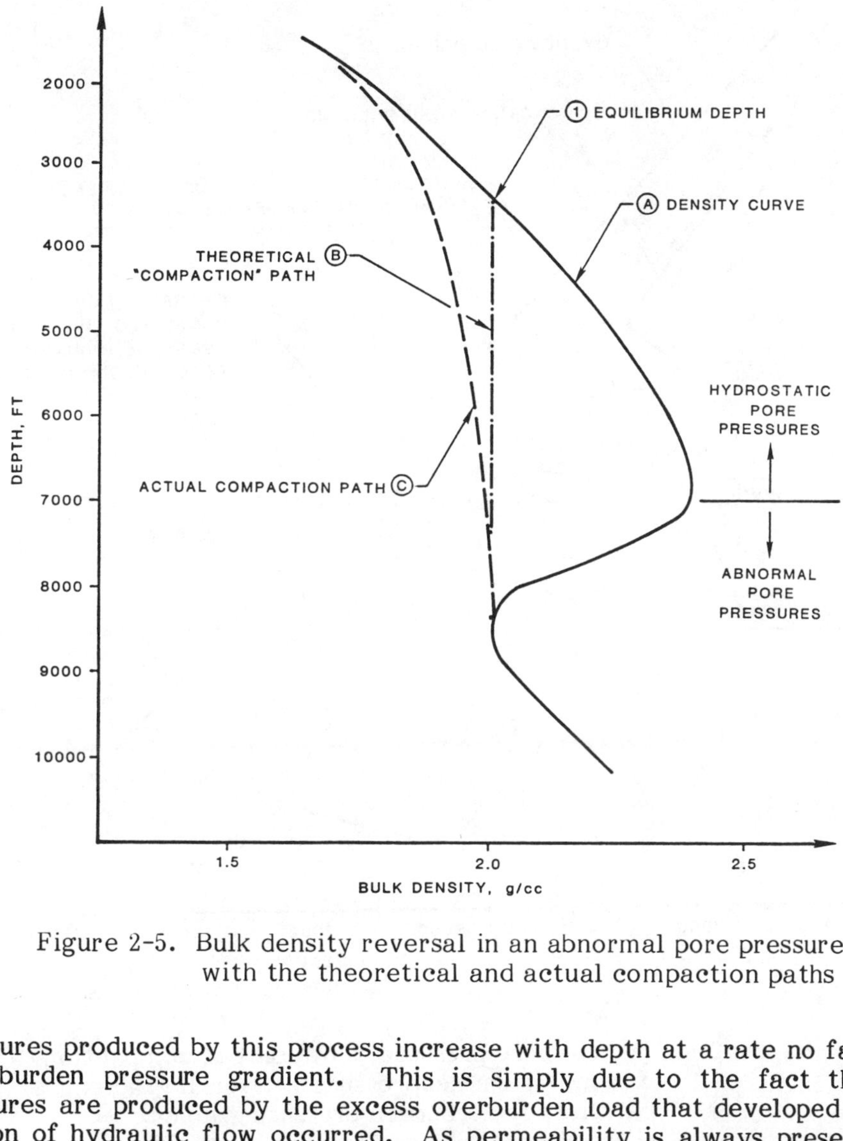


Figure 2-5. Bulk density reversal in an abnormal pore pressure zone, with the theoretical and actual compaction paths

Geopressures produced by this process increase with depth at a rate no faster than the overburden pressure gradient. This is simply due to the fact that these geopressures are produced by the excess overburden load that developed after the restriction of hydraulic flow occurred. As permeability is always present (albeit extremely small in clay rocks), it is normal for some leakage to occur, especially when the hydraulic potential becomes great. Thus, over a large interval, the pressure gradient is normally less than the overburden pressure gradient. If the geopressure zone contains some thin permeable sandstone beds which possess a hydraulic conduit to a lower potential, flow will occur from the adjacent clays to the sand. In this manner the clay will be able to compact adjacent to the sand aquifers, which will cause a decrease in porosity and permeability of those clays and thus restrict further flow. The pressure gradient across these clay/sand/clay zones will be significantly greater than the overall pressure gradients, but it must be remembered that the local modifications of the pressure gradient across the permeable zones are due to leakage, and the pressure gradient will be a function of the permeability of the adjacent clays and the magnitude of the hydraulic potential across the clay/sand boundary. Figure 2-6 illustrates the above phenomenon as a function of pore pressure.

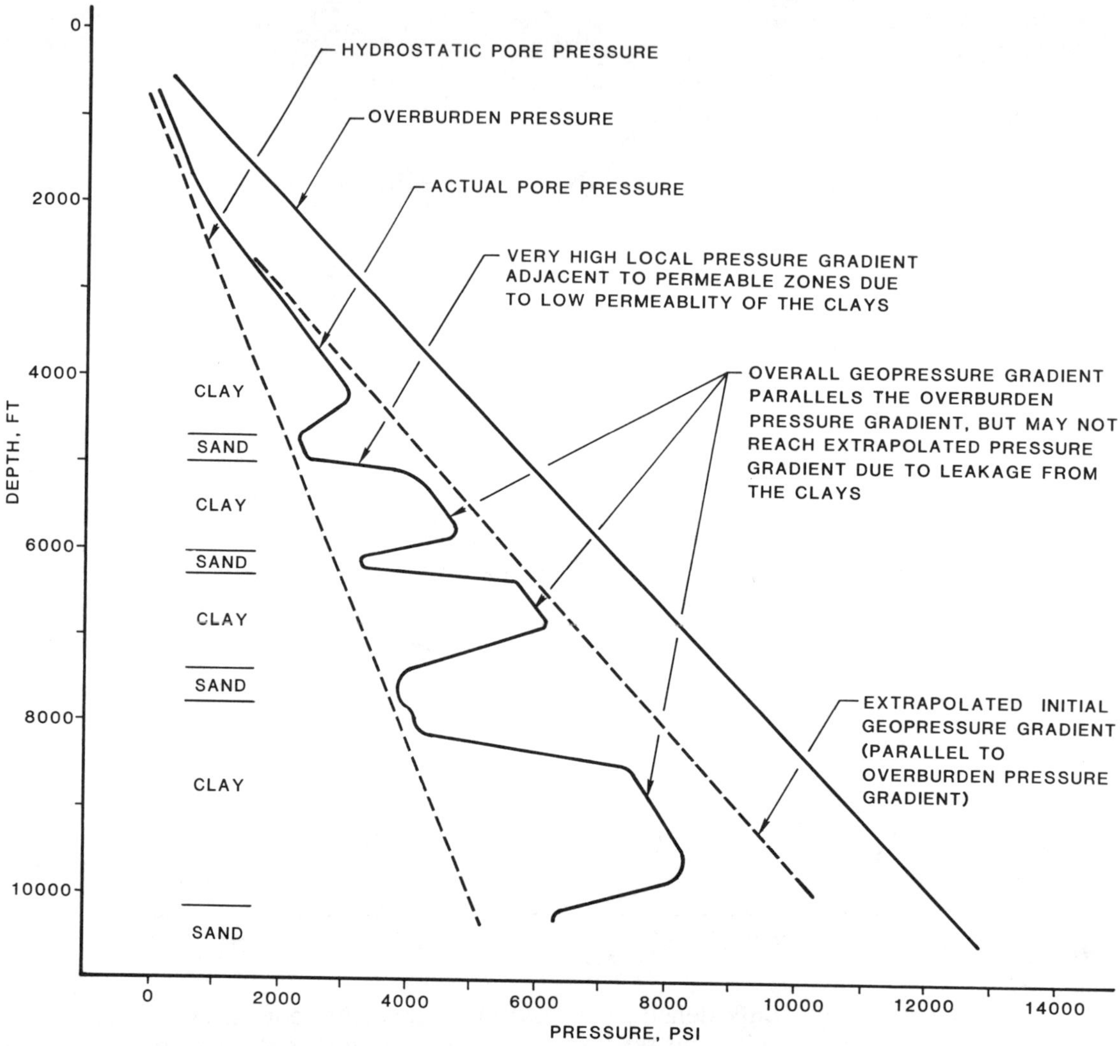


Figure 2-6. Typical pore pressure-depth plot of compaction disequilibrium geopressures. Overall gradient is parallel to overburden pressure gradient

## 2.3 AQUATHERMAL PRESSURING

The effect of temperature on pore pressure was originally described by Barker in 1972. The basis for Barker's hypothesis is a pressure-temperature-density diagram for water (Figure 2-7) which was adapted from a pressure-temperature-volume diagram for water and carbon dioxide (Kennedy and Holser, 1966). In using this diagram for determining pressure changes with temperature in a subsurface environment, Barker makes the following provisos:

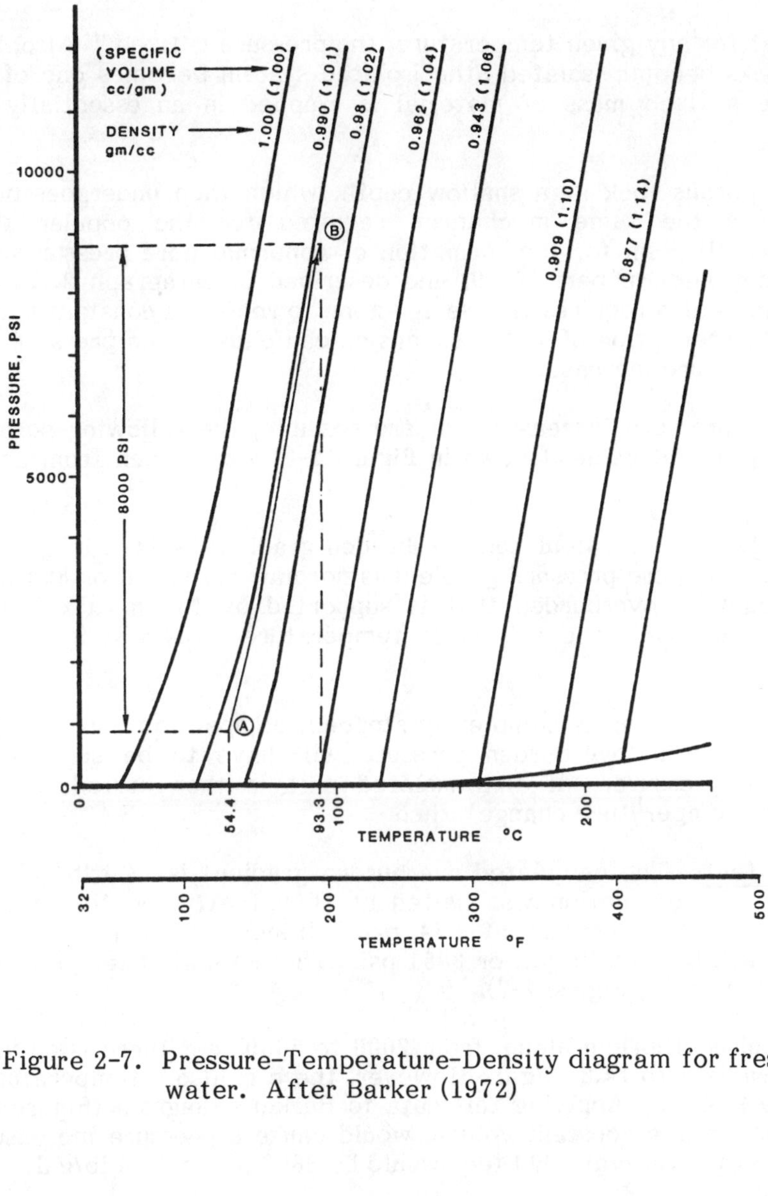


Figure 2-7. Pressure-Temperature-Density diagram for fresh water. After Barker (1972)

1. **Porous volumes must be completely isolated from their surroundings.**

2. **The porous volume that will become geopressured upon a temperature increase must have a constant volume.**

3. **For geopressure zones, isolation must have occurred at a lower temperature than those observed now.**

**Justification for use of this diagram is provided by Barker's statement: "Because depth and pressure are related, the pressure and temperature of the formation fluids which follow the normal hydrostatic gradient are related by the geothermal**

gradient, so that for any given temperature, the pressure is fixed." Also, that "...if some of the rocks become isolated...the isolated system becomes one of constant density because a fixed mass of material is trapped in an essentially constant volume."

This sealing of porous rock at a shallow depth, which then undergoes burial to a greater depth, is the same mechanism required for the popular theory of compaction disequilibrium for the formation of abnormal pore pressures, first put forward by Rubey and Hubbert (1959) and described in paragraph 2.1. However, Barker's hypothesis also requires this sealed zone to remain a constant volume with burial so that the prevention of water expansion can cause rapid pressure increases with small temperature increase.

To illustrate the pressure increases that are possible, the following pore pressure and overburden gradient values (shown in Figure 2-8) were taken from an offshore well:

> At 2000 feet: The calculated overburden gradient is 15.0 lb/gal, or 1557 psi. The pore pressure gradient is normal, 8.6 lb/gal or 893 psi. The amount of overburden that is supported by the matrix is 6.4 lb/gal or 664 psi. The estimated temperature is 54.4$^{\circ}$C (point A, Figure 2-6).

If at this depth dewatering is completely stopped, compaction cannot proceed, so any further increase in overburden pressure will have to be supported by an increase in pore pressure. If this isolated unit is then buried to 5000 feet (disregarding any temperature change), then

> At 5000 feet: The calculated overburden gradient is 17.4 lb/gal, or 4515 psi. If compaction was halted at 2000 feet, then the matrix stress remains unchanged at 664 psi. Hence, the pore pressure gradient will be 14.8 lb/gal, or 3851 psi. The estimated temperature is 93.3$^{\circ}$C (point B, Figure 2-7).

By compaction disequilibrium alone, from 2000 to 5000 feet there is an increase in pore pressure gradient to 14.8 lb/gal. However, there is also a temperature change from 54.4$^{\circ}$C to 93.3$^{\circ}$C. Applying this data to Barker's diagram (Figure 2-7), this temperature change in a constant volume would cause a pressure increase of 8000 psi. The total net pressure at 5000 feet would be 8893 psi or 34.3 lb/gal.

The actual rate of pressure increase in this example is 2.67 psi/ft, corresponding to a geothermal gradient of 1.3$^{\circ}$C/100 ft. The overburden would be balanced by the pore pressure at approximately 2500 ft, where the temperature is 64.1$^{\circ}$C.

The change in volume of water due to a temperature increase from 54.4$^{\circ}$C to 64.1$^{\circ}$C is approximately 0.44 percent, and an increase of only 0.39 percent is required to obtain the pressure increase needed to balance the overburden pressure at 2500 ft. Barker states that the pressure-temperature-density diagrams for NaCl solutions closely resemble the diagram for water, but that the excess pressure does not rise as rapidly with temperature increase for a saline solution as for pure water.

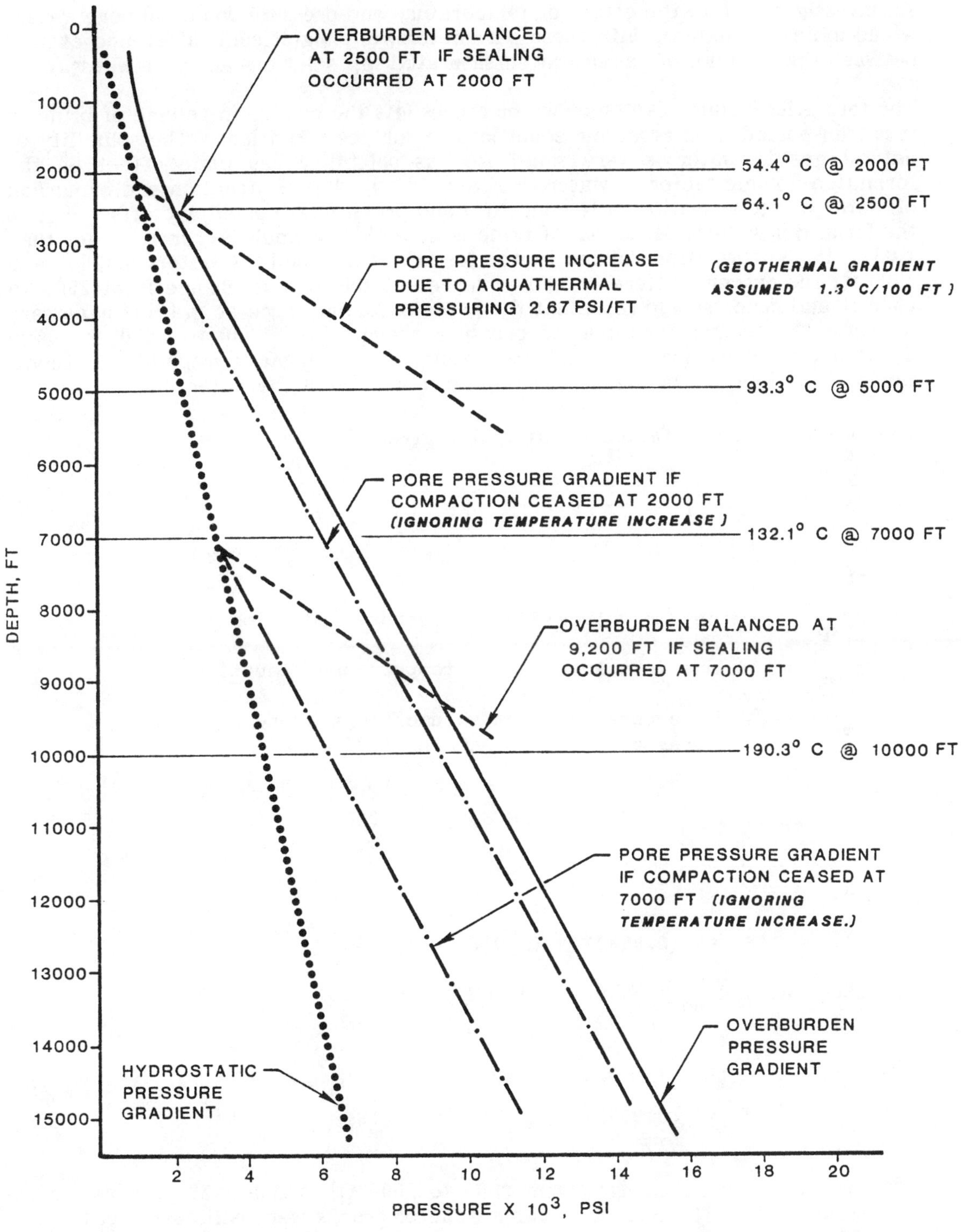


Figure 2-8. Pore pressure increases with aquathermal effect

An investigation into the effect of temperature and pressure on actual pore waters would indicate whether, with these parameters, any significant difference existed between the behavior of gas saturated/saturated salt solutions and pure water.

The formation-volume factor of water represents the change in volume of brine as it is transported from reservoir conditions to surface conditions. The solubility of natural gas in water is very small, so gas solubility has little effect on the formation-volume factor of water (McCain, 1973). The contraction and expansion of water due to reduction in temperature and pressure are small and offsetting, so the formation-volume factor (B) of brine is numerically small — rarely larger than 1.06. If the formation-volume factor for brines is found for waters at 2000 and 2500 ft, the percent difference should represent the volume change from 2500 to 2000 ft and hence should represent the percent volume increase in brine necessary to cause the temperature-induced pressure change from 2000 to 2500 ft. Any difference between this value and that obtained from Magara's graph (0.39 percent) would be attributable to the difference in composition of the water.

The formation-volume factor of water, B, is given by:

$$B = (1 + dV_{wp}) * (1 + dV_{wt}) \tag{2-1}$$

where

$dV_{wp}$ = change in volume due to pressure reduction

$dV_{wt}$ = change in volume due to temperature reduction

$dV_{wp}$ and $dV_{wt}$ are obtained from volume/temperature/pressure graphs, McCain (1973).

At 2000 feet, $dV_{wp} = -0.0003$ and $dV_{wt} = 0.016$

hence,

$$B = (0.9997) * (1.016) = 1.0157$$

At 2500 feet, $dV_{wp} = -0.0025$ and $dV_{wt} = 0.021$

hence,

$$B = (0.9975) * (1.021) = 1.0184$$

The percent volume change (from 2500 to 2000 ft) is thus 0.27 percent for gas-saturated brine. The formation-volume factor graphs were calibrated for brines of 300,000 ppm NaCl, and saturated with methane. Hence for waters of this composition, the expansion due to temperature increase of 9.7°C (from 54.4°C to 64.1°C at 2500 ft), and contraction due to pressure increase (893 psi at 2000 ft to overburden pressures: 2,130 psi at 2500 ft) of a volume that is buried from 2000 to 2500 ft, are only half that which is predicted for pure water. The majority of formation pore waters falls somewhere between these two extremes of composition, so a reasonable value for volume change would be approximately 0.3 percent for the above parameters.

With this modification, if a sealed volume at 2000 ft undergoes burial, the pore pressure increase due to pore water expansion would produce overburden pressure values at approximately 2900 ft. This reduction in rate of expansion with temperature does not significantly detract from the overall mechanism: the rate of pressure increase is still extremely rapid, providing isolation is total.

If a zone of porous rock could be sealed totally from its surroundings in nature, during burial and earth movements the coherence of the seal would probably exist for a very short period of geological time. Rocks of zero permeability (such as salt, evaporites and some limestones), if they are not fractured, could provide such sealing. As Barker mentions, shales are permeable, and the volume of water that would need to be released in order to reduce the temperature-induced pressure is very small. Magara (1975, a), in explaining the average 1.4 psi/ft gradient for pore pressure increase for Gulf Coast wells, notes that there was possibly no perfect isolation of fluids in the undercompacted shales, as the 0.76°C/100 ft geothermal gradient should produce aquathermal pressure gradients of approximately 1.8 psi/ft. An explanation involving the expansion of the pressured formation to increase the volume and hence reduce the pressure is also feasible, but Magara (1975, b), feels this is unreasonable as expansion of rock due to any mechanism (be it thermal expansion of water or montmorillonite dehydration) is not easy to explain geologically (paragraph 2.4). However, no evidence has been put forward to indicate that expansion of sediment does not occur.

In summary, a temperature increase causes pore waters to expand approximately 300 times more than a typical rock matrix, and if a sealed volume undergoes a temperature increase through burial, very rapid pressure increases can occur. If abnormal pore pressures were produced by aquathermal pressuring, what effect would the mechanism have on current wellsite pressure evaluation techniques?

If by any mechanism a porous volume became completely isolated and pressured, then no transition zone would exist (see paragraph 2.10). As mentioned above, the seal must be totally impermeable. Also, if the seal remained intact during burial, then the maximum pressure developed within the porous rock would be equal to the overburden pressure, plus the tensile strength of the seal. Fertl (1976) states that pore pressures exceeding overburden pressures by up to 40 percent are known. In these cases burial must be shallow, as laboratory experiments have shown that the tensile strengths of sedimentary rocks rarely exceed 1000 psi. Detection of these pressures would be severely limited with current methods. Probably the only diagnostic tool would be flowline temperature. If such a zone of abnormal pressure existed within an impermeable rock, heatflow would be severely disturbed. Lewis and Rose (1970) showed that water is a threefold greater insulator than most sedimentary rocks, except shale, so the sealed volume would be a very effective insulator. Hence, if penetration of an impermeable zone is accompanied by a steady significant decrease in flowline temperature, it is possible that a porous, high-pressure zone is being approached.

If the porous zone is "sealed" by a thickness of shale, then a transition zone exists and most of the normal pressure "detection" techniques can be applied. The transition zone is indicative of permeability within the shale.

Bradley (1975) and Magara (1975) confirm that the volume of water which produces excess pressure is extremely small, and release of such volumes over short periods of geologic time is well within experimental ranges of hydraulic conductivities and observed permeabilities of shales. If the flow of water from high hydraulic potential to low hydraulic potential is faster than or equal to the rates of pressure build-up, then there is no net pressure gradient increase (see paragraph 3.5). However, if the rate of pressure build-up is extremely rapid (as in aquathermal pressuring), it is possible that the hydraulic conductivity of shale is exceeded, resulting in a net pressure gradient increase.

With both a fast burial and a high geothermal gradient, probably only a short period of geologic time — i.e., that required for burial from 2000 to 2900 feet — will elapse before the pressure in the sealed volume could exceed the overburden gradient. Upon further burial and temperature increase, the generated pressure would fracture the "seal" and release the excess water. As shown by Magara (1975), the amount of water released would be equal to the amount of volume expansion caused by temperature increase plus the volume causing excess pressure due to subcompaction.

It appears that aquathermal pressuring would be more important as a mechanism for generating abnormal pore pressures at relatively shallow depths where the rate of burial and compaction is generally faster than at greater depths. However, due to the excessive rate of pressure build-up predicted by Barker's diagram, it would seem that abnormal pressures generated in this manner would be geologically transient affairs, as illustrated by the above example.

If pressure is sufficiently contained within a porous unit (i.e. a vuggy limestone) and expansion occurs which is sufficient to release the excess pressure through the formation of fractures, then the flushing action may carry some hydrocarbons contained in the water to zones of lower hydraulic potential. As the volume of water released is small, this mechanism probably is not a major force instrumental in the migration of hydrocarbons, but the formation of hydraulic conduits may well be extremely important in providing pathways for later hydrocarbon movement.

Barker shows that if trapping occurs at 12,500 feet, then aquathermally-generated pressures can exceed the overburden pressure at geologically reasonable depths. The possibility of a perfect seal forming at such depths and retaining its coherence to 20,000 feet is probably remote. Much of the plasticity of sediments at these depths has been removed by compaction and diagenesis, such that fissility and natural fractures formed through these processes would effectively subdivide a seal, allowing pressured fluids to escape.

It would appear that aquathermal pressuring may be a mechanism through which abnormally high pore pressures are formed at shallow depths. Should the seal remain coherent with burial, then with further temperature increase the release of excess pore pressure could be a major process in the formation of natural fractures and migrating hydrocarbon conduits. However, the maintenance of constant, sealed volume with the high pressures which would rapidly develop with burial is unlikely. It is probably that aquathermal geopressures are transitory phenomena.

On the other hand it is possible that the thermal expansion of fluids may play a role in the creation and maintenance of permeability and hence the prevention or depletion of geopressures (see paragraph 2.11). Rock fabric thus created may play a later role in the migration of water and hydrocarbons.

## 2.4 MONTMORILLONITE DEHYDRATION

Since 1950 it has been known that, with burial, montmorillonite (or more correctly, smectite) alters to become illite. The process was originally proposed to explain the abundance of montmorillonite at shallow depths and the abundance of illites at greater depths. Subsequently, geologists (Powers, Burst and Weaver, 1959-1961) found that the transition from montmorillonite to illite is depth-related but is also strongly affected by temperature and ionic activity. Thus it appeared that illite is formed from montmorillonite by normal diagenetic processes. However, within a small depth range, clay alteration zones have been found which cross complete chronological sections. Thus it is now generally accepted that, although montmorillonite diagenesis is not related to geologic age, in some areas it may have this appearance because geologically older sediments usually are buried deeper than younger rocks. The rate of alteration appears to be mainly temperature-dependent such that, in areas of high geothermal gradients, alteration occurs at a shallower depth than in those areas with low geothermal gradients.

The name "montmorillonite" was originally applied to a clay mineral with composition similar to that of pyrophyllite except for the presence of excess water. The general formula for smectites is

$$Al_4\,Si_8\,O_{20}\,(OH)_4 * nH_2O$$

Chemical variations of this basic formula yield a group of clay minerals which contain beidellite, hectorite, montmorillonite, nontronite, saponite and sauconite. All have the capability of adsorbing water or organic liquids between their structural layers, and all exhibit marked cation exchange properties. Montmorillonite (a sodium/magnesium variety) and beidellite are the principle constituents of bentonite clay deposits. (Bentonites were formed by the alteration of eruptive igneous rocks, largely by halmyrolitic processes.)

Figure 2-9 illustrates the changes in ion substitution from one basic pyrophyllite sheet through smectite to mica. The basic structure of these phyllosilicates (Greek: phyllon, leaf) arises from the dominance in structure of the silicon-oxygen sheet. Three of the four oxygens in each $SiO_4$ tetrahedron are shared with neighboring tetrahedra. This structural unit is sometimes called the "siloxane sheet". The simplest type is pyrophyllite which is made up of two $Si_2O_5$ sheets either side of an octahedral sheet coordinated by hydroxyl and aluminum ions. The net charge is zero, and the forces holding the units together are the van der Waal's bond: since this is very weak, the mineral has a soapy feel. Substitution of aluminum ions for some silicon ions causes a charge imbalance as aluminum is trivalent, whereas silicon is tetravalent. The charge imbalance causes adsorption of cations and water between the units, and this is characteristic of the smectite group. Further substitution of aluminum for silicon increases the charge imbalance, and in the presence of potassium or sodium ions which become firmly fixed in the interlayer sites, the charge becomes neutralized, and the mineral loses its soapy texture to become fissile. The illite group has random substitution of Al for Si, whereas the mica end-member has aluminum substituted for every fourth silicon.

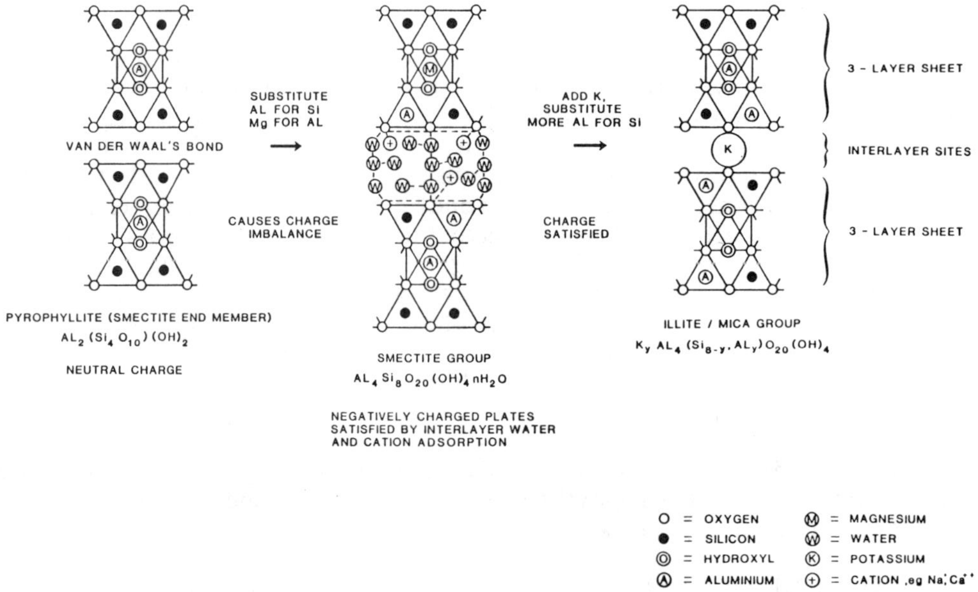


Figure 2-9. Changes in ionic substitution in three-layered sheets

Illites are more closely related to the micas, and they have the general formula:

$$K_y\ Al_4\ (Si_{8-y},\ Al_y)\ O_{20}\ (OH)_4$$

where y is usually between 1.0 and 1.5. Since pure illites generally contain little or no interlayer water and are not penetrated by organic liquids, they show no swelling characteristics. The presence of interlayer potassium ions prevents the adsorption of water or hydrocarbons and other cations, hence illites have a low cation exchange capacity.

The charge imbalance in smectites causes water to be adsorbed between the lattice units. Although the water molecule is electrically neutral overall, it has uneven electron distribution — causing the hydrogen to have a slight positive charge. Thus water becomes hydrogen-bonded (Figure 2-10) to the sheets, and cations are loosely held to the oxygens. The bonds are weak, except for those immediately adjacent to the clay sheets, allowing cation exchange.

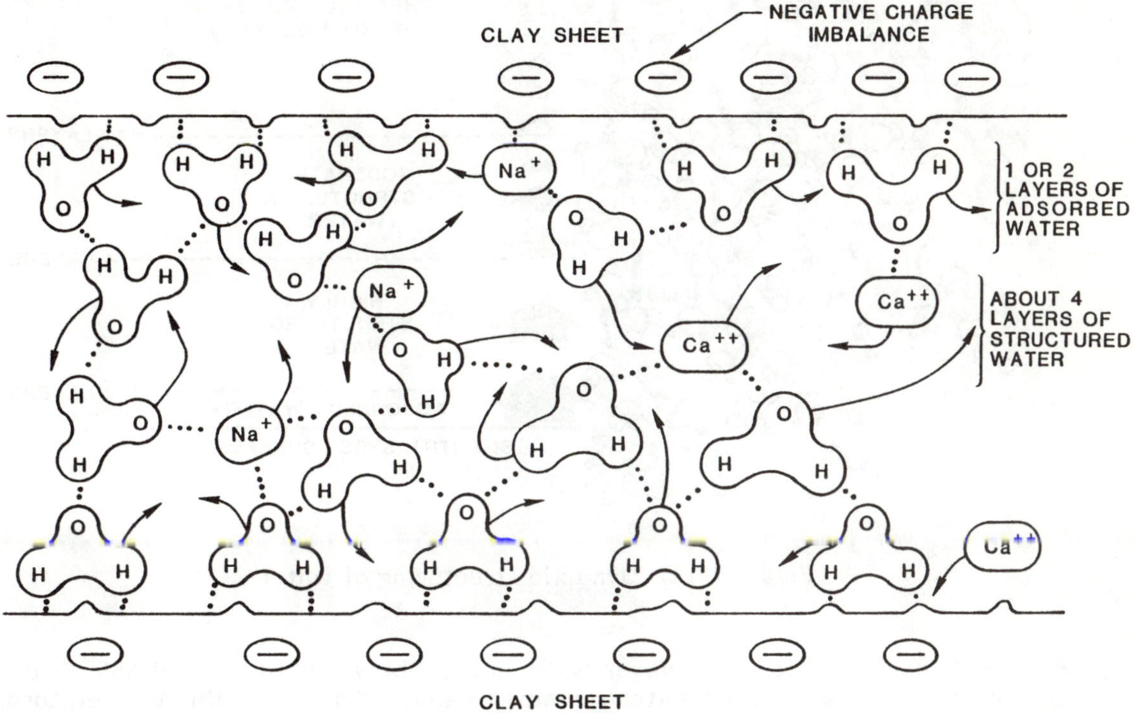


Figure 2-10. Hydrogen-bonded water and exchangeable cations

The amount of interlayer water adsorbed by smectites varies according to the type, the nature of the interlayer cations and the prevalent physical conditions. Calcium smectites usually take up two layers, while sodium varieties take up considerably more. Powers (1967) indicates that at least four monomolecular layers are adsorbed on interlayer sites, and up to ten layers are held on the outside of the plates (Figure 2-11). The inner four layers are highly structured and are packed in a hexagonal form. This packing imparts a rigidity to the water such that its viscosity is similar to that of ice. Also, it has been suggested that the density of this highly structured water is greater than 1 g/cc, and estimates vary from 1.4 to 1.7 g/cc.

Most of the water adsorbed onto the surface of illite particles and the small amount that may be interlayered with illite sheets is liberated rapidly below 110°C, and the remainder is expelled more slowly between 110°C and 350°C.

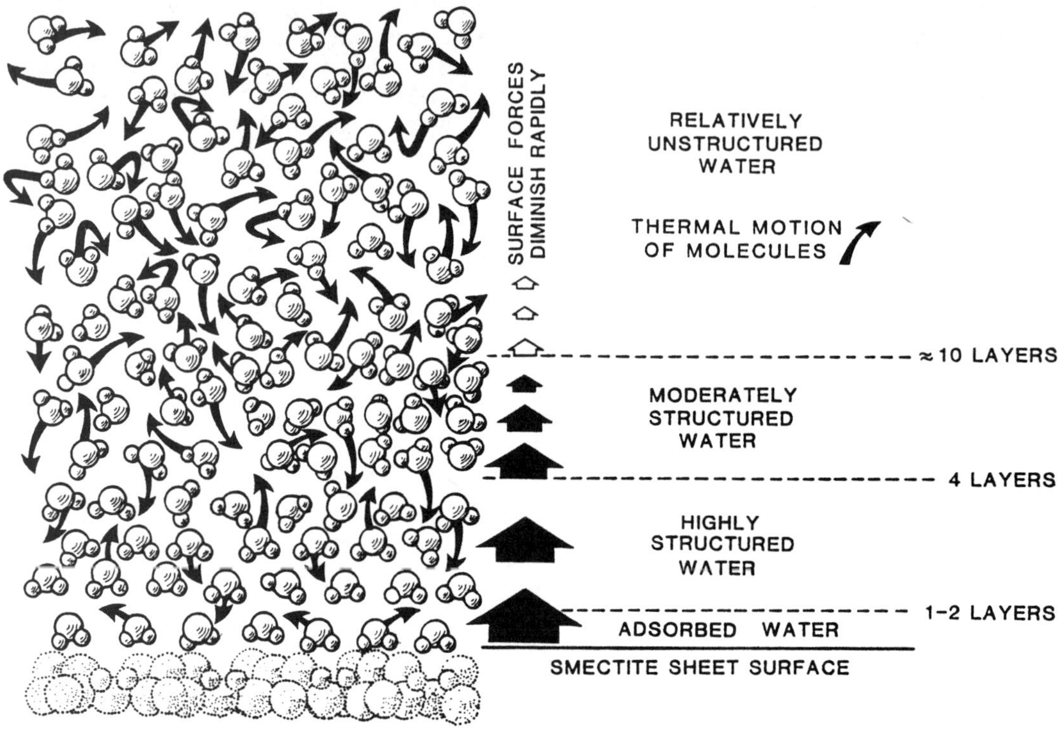


Figure 2-11. Dynamic structuring of water

With heat, smectites lose interlayer water mostly between 100°C and 250°C, but some of the more structured water remains to about 300°C. At this temperature, slow loss of constituent (OH) begins.

There is considerable controversy among petroleum geologists as to the significance of (1) the amount of adsorbed water on clays, (2) the density of structured water, and (3) the temperature range at which dehydration is initiated. Some workers (Hinch, 1978; Magara, 1975; and others) believe clay dehydration is strongly instrumental in the redistribution of protohydrocarbons within the sediment, and not necessarily a major force in the generation of abnormal pore pressures. Most petroleum and some clay geologists suggest that initial dehydration may occur at temperatures as low as 60°C.

The implications of this thermal dewatering of expandable clays are mainly that (1) not only does pore water have to be expelled for compaction to occur, but that (2) additionally, huge volumes of adsorbed water must be released in order that diagenesis may progress. Experimental compaction of clays has shown that pressure plays a minor role in the reduction of porosity. Heat, ionic activity and hydraulic flowpaths are the major forces in clay diagenesis and compaction. Temperature causes dewatering and aids the collapse of expanded smectite layers, and the adsorption of potassium ions completes the transformation to illite. The presence of potassium ions at temperatures lower than that necessary for dewatering also causes collapse of the smectite lattice, and thus illite may be formed at low temperatures.

In order to suggest that alteration may be occurring at a certain depth, it must be assumed that the sediment source and other geological parameters have remained constant during deposition and diagenesis of the sediment. Illites may have been deposited after their formation by weathering of silicates (principally feldspars), or they may be derived by diagenetic alteration of other clay minerals. Smectites, other than those produced by the alteration of extrusive igneous rocks, can be produced by weathering of basic igneous rocks, and also may occur as hydrothermal alteration products. Montmorillonite forms when sufficient magnesium is present. Alteration of basic rocks yields illites unless magnesium and calcium ions are plentiful and potassium is absent or low in concentration (Deer et al, 1966).

Thus the consensus is that montmorillonite becomes dehydrated and partially collapsed to metamontmorillonite or degraded illite; then, with further temperature increase and potassium substitution, all the adsorbed lattice water is driven off and true illite is formed.

Powers (1967) suggested that, as the last four adsorbed layers of monomolecular hydrogen-bonded water may have a density of approximately 1.4 g/cc, upon desorption of the water to the pore spaces, the "phase change" will also be accompanied by a density change — the structured water will expand until the density has decreased to approximately 1 g/cc. If the pore spaces are already filled with pore water, this relatively sudden influx and expansion of desorbed water creates a space problem such that an expansion of the order of 20 percent of the bulk porosity would be necessary to accommodate this water. Clearly an expansion or rebound of a clay zone of this order would involve considerable disruption of the overlying sediments. Fluid pressure buildup would depend on both the formation rate of normal density water and the continuing simultaneous escape of pore water from the clay body. Magara (1975) questioned Powers' figures and showed that a probable maximum rebound of a clay body would be around 6 percent; however, even this volume increase is insufficient to explain the bulk density reduction that is consistently measured within a geopressured clay. Also, expansion or rebound of a clay body would be exceedingly difficult to compromise geologically: the large volumes of released structured water would cause an extremely rapid pressure increase if the mechanism occurs as postulated.

Figure 2-12 shows hypothetical water escape curves with depth from a montmorillonite section. Burial depth may be equated to temperature if the geothermal gradient is known; thus if geopressures do occur due to this mechanism, they will be encountered below the 60°C to 100°C isotherms. Refer to Section 4 for qualitative geopressure evaluation techniques.

In summary, the process of montmorillonite dehydration and its alteration to illite with the concomitant formation of geopressures is as follows.

Sedimentary smectite, during transport and shallow burial, becomes hydrated and adsorbs up to ten monomolecular hydrogen-bonded water interlayers. At this stage the lattice is expanded to its maximum and the environment would be alkaline, rich in calcium and magnesium, and poor in potassium. After further burial, compaction causes expulsion of free pore water such that the "house of cards" clay structure becomes compacted to a "herringbone" type structure. Further burial and temperature increase cause all but the last (approximately) four layers of

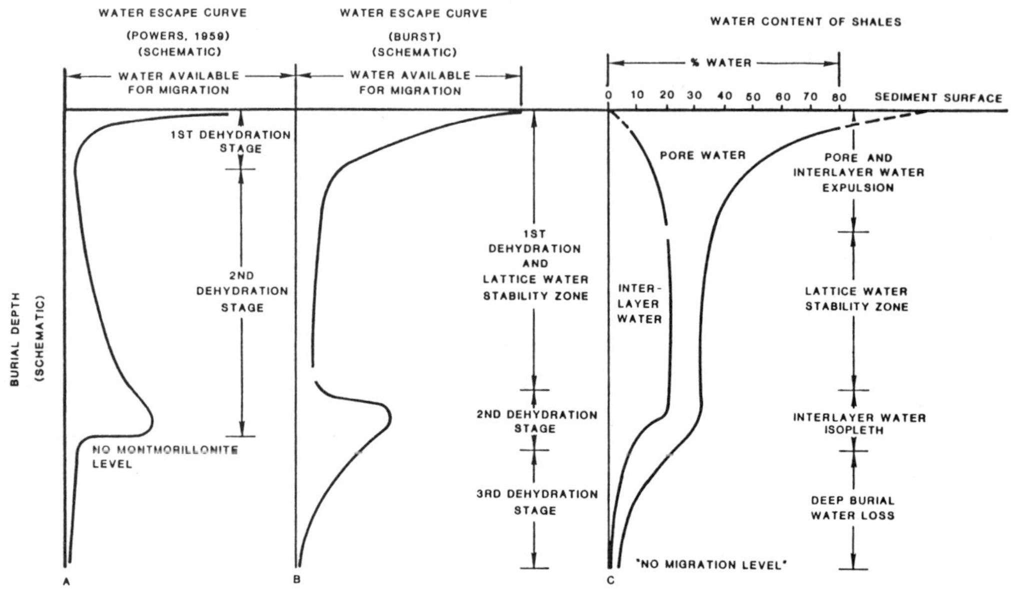


Figure 2-12. Hypothetical dehydration curves of montmorillonite sediments with depth and temperature

structured water to be desorbed and liberated to the pore spaces. The rock at this stage would be fairly well compacted and may develop fissility. Figure 2-13 is a schematic of compaction and alteration stages. As burial progresses through the 100°C isotherm, structured water becomes liberated to the pores. If when desorbed this water expands due to a density reduction from the structured phase to the pore phase, pore pressure will increase. As interlayer water is released, the clay lattice will collapse. With higher temperatures, all interlayer water except the last layer will be desorbed and the clay lattice can collapse further: the amount of water released to the pores is volumetrically greater than the volume reduction caused by lattice collapse. With the availability of potassium, diagenesis to illite occurs. This reaction involves adsorption of potassium at the interlayer and surface sites as well as the release of a small amount of silica. Upon completion of this process, the diagenetic alteration of smectite to illite is complete.

Thus if this process is instrumental in producing geopressures, the rate of expulsion of free pore water with burial has to be less than the rate of production of desorbed water from the clay interlayers. If potassium is scarce, alteration to illite will not be completed and the resulting clay will be a mixed layer type or degraded illite. Also, if drainage is restricted after diagenetic alteration to illite, the silica produced from the process will be precipitated in the pore spaces. The significance of this precipitated silica in further reducing permeability and as a cementing agent to adjacent aquifiers is conjectural. Most authors consider it to be a secondary phenomena but some claim it to have major significance in permeability reduction and geopressure development.

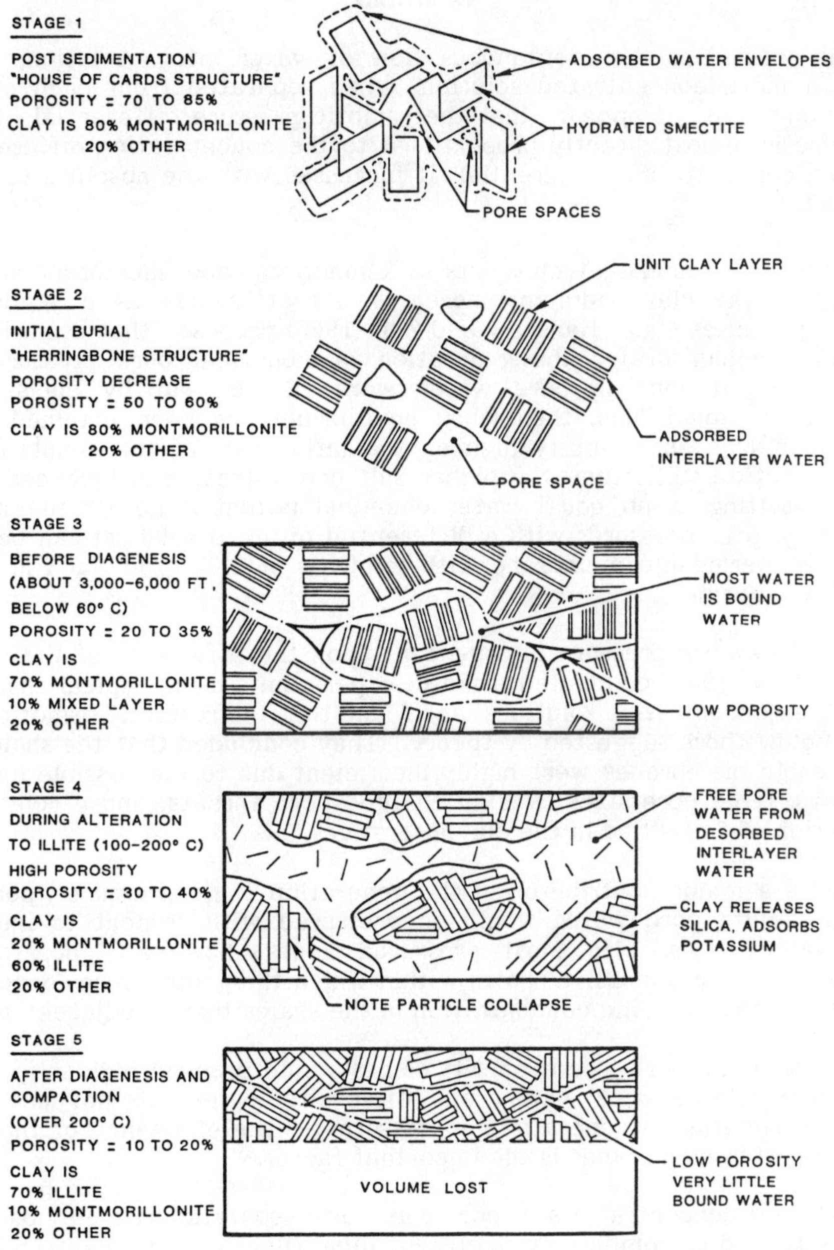


Figure 2-13. Diagenetic stages in the alteration of montmorillonite to illite

## 2.5 OSMOSIS

Osmosis is defined as the spontaneous flow of water into a solution, or from a dilute to a more concentrated solution, when separated from each other by a suitable membrane. It appears that the osmotic pressure differential at constant temperature is almost directly proportional to the concentration differential; and for a given concentration differential, it increases with the absolute temperature (Jones, 1969).

Researchers have found that clay acts as a semipermeable membrane and that, as the purity of the clay sediment increases, its efficiency as a semipermeable membrane increases also. Hanshaw and Zen (1965) proposed that osmosis might be a plausible mechanism for the generation of abnormal pore pressures. They suggested that if ions in pore water were filtered out by clays acting as semipermeable membranes, then when equilibrium has been attained, the pore pressure should be anomalously high on the influx side of the membrane. Upon reaching osmotic equilibrium, the higher salt concentration is balanced by higher pressure, resulting in an equal water chemical potential across the membrane. Theoretically, pore pressures with a differential of up to 4500 psi can be produced across a semipermeable membrane with solutions of 1.02 g/cc NaCl in water and saturated NaCl brine.

A generated osmotic pressure is dependent upon the efficiency of the membrane. Young and Low (1965) conducted osmotic experiments using typical shale samples and variable concentration solutions, and found that the resulting osmotic pressures were far below those suggested by theory. They concluded that the shales used as semipermeable membranes were highly inefficient due to the possible existence of microcracks, large pore sizes, weakly charged pore throats, and a relatively high concentration of fine silica in the clay interstices.

If osmosis is a major contributor in the generation of geopressures, porous zones with high salinity pore water would be pressured with respect to impermeable fresher water zones. In most pressured zones, however, geopressures are associated with thick, massive shales within which the geopressures occur, and the pore water is lower in ionic concentration in the shales than in adjacent sandstones.

The presence of evaporite beds or salt domes would produce high salinities in the adjacent and related pore water, but these lithologies are not necessary for establishing osmotically generated geopressures — the efficiency of the clay beds as semipermeable membranes is the important factor.

If two different concentrations of pore water are separated by a clay bed and their compositions tend to combine in electrochemical reactions, an electrical potential develops across the clay layer and induces liquid movement called electro-osmosis. Also, a temperature difference between the two concentrations tends to cause liquid movement known as thermo-osmosis. Clay mineralogists have found that if the osmotic, electro-osmotic and thermal-osmotic conductivities combine they can be at least as great as the hydraulic conductivity, and their importance increases with compaction or decreasing porosity. It is thus apparent that, due to a pressure differential, water may flow through a clay bed; but due to salinity, heat and chemical composition of the opposing pore fluids, hydraulic flow may cease or even be reversed. In any case, the effective permeability of the clay bed determines the maximum rate of water movement.

Although electro-osmosis and thermal-osmosis have been displayed in the laboratory under ideal conditions, to date no evidence has been put forward that these processes occur in a subsurface environment. However, electro-osmosis has been shown to have appreciable effect in a simulated subsurface environment (Olsen, 1972).

In summary, if osmosis is to be a major geopressure-generating mechanism, a clay bed separating two sandstone beds (having different salinity pore fluids) must be an efficient semipermeable membrane. Water will then move from the low concentration through the clay to the higher concentration. If hydraulic conductivity is restricted in the latter sand body, the pore pressure will rise.

## 2.6 REVERSE OSMOSIS

Reverse osmosis may occur across a clay membrane whenever pressure is applied on the more saline side, causing water to migrate to the sand bed which has the less saline pore water. Hence in a sand/shale sequence with overall pore water freshening with depth, as in the Gulf Coast, this hypothetical process would not be operative because the gravitational load is higher in the less saline deeper zones. Reverse osmosis has been better termed "diffusion pressure" by Rieke and Chilingarian as it describes the process more accurately.

## 2.7 IONIC FILTRATION

The process of ionic filtration in the subsurface environment is not thought to be a mechanism that produces geopressures; however, a description of the process should aid understanding of the possible origin of varying salinities with different lithologies.

Normally, the concentration of dissolved ions within pore water increases with depth. Seawater, trapped in sediments during deposition, might have been diagenetically changed to assume its present chemical character and concentration within any formation. Indications are that some physical process is actively concentrating ions within permeable formations — either those ions of the original water of deposition, or the ions dissolved in meteoric water entering at the outcrop. Experimental studies (Brederhoeft et al, 1963) show that clay acts as a membrane that restricts the passage of ions as water passes through the clay.

Membranes composed of certain clays are permeable to cations but not to anions. Specific types of clay particles within membranes are negatively charged and therefore repel the negatively charged anions. Water molecules that do not ionize to the same extent as salts pass through the membrane in response to hydrostatic potential. The anions do not pass; they attract cations, thus preventing cation passage also. If the hydraulic gradient tends to move water through argillaceous, less permeable beds which have a membrane effect, ion concentration occurs in the more permeable formation.

## 2.8 TECTONISM

Several geopressure occurrences are in areas of intense post-Miocene or post-Pliocene deformation. The Qum field in N.W. Iran has exceedingly high geopressures in a Tertiary limestone just below a small post-Pliocene bedding plane thrust fault. Also, the Aghar Jari field in west-central Iran has geopressures associated with steep folding and small thrust faults; the Khaur field in West Pakistan, the Tupungato field in Argentina, the Ventura Avenue field in California, and widespread geopressures in Trinidad are all thought to be intimately associated with active or past tectonic activity, to cite a few.

The mechanism of generation of geopressures due to tectonic compression is very similar to the process involved in compaction disequilibrium (paragraph 2.2) with the exception that the time factor is a function of strain rate and not rate of deposition.

A useful measure of the degree of compaction of a clay is its porosity. (Porosity is defined as a ratio of the pore volume to the total volume.) The porosity in clay can be expressed as a function of the amount of stress that the clay can support without further compaction. Thus in nontectonic geosynclinal basins, porosity may be directly related to the effective weight of the overlying sediments if the pore water can escape in response to increasing overburden pressures. However, in a tectonic environment, additional compacting force (tectonic stress) is applied horizontally. If pore water can escape sufficiently fast to maintain equilibrium at all times, then normal pore pressures will prevail; however, the clay will be denser in the tectonic environment at a particular depth due to the reduction in porosity caused by the tectonic stress.

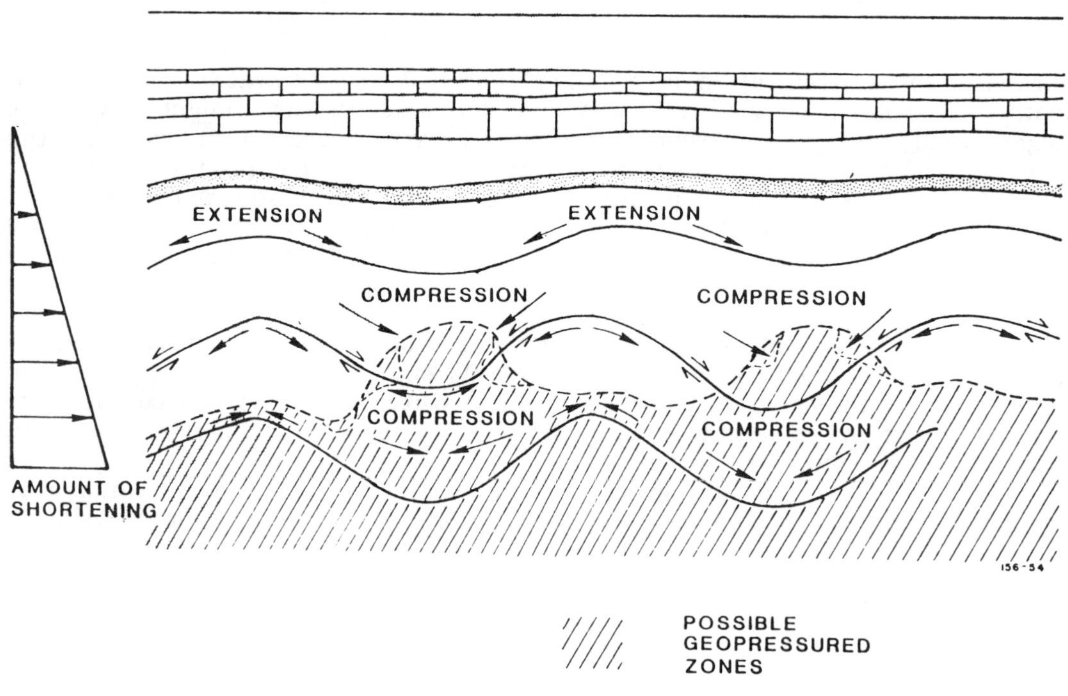


Figure 2-14. Geopressures caused by tectonic compressional folding

As tectonic load is added, the immediate effect on the clay strata is for the increment of load to be assumed by the pore fluid. Then as the excess water is slowly expelled from the clay, compaction occurs due to the concomitant decrease in porosity and the increase of effective stress in the solid. Since it is likely that the rates of overburden stress buildup and expulsion of pore water from compacting clays are very close (as indicated by the occurrence of geopressures that are generated due to rapid deposition), it would appear that any further increase in stress could cause disequilibrium: pore water would not be able to escape at a rate equal to the volumetric reduction of pore space, resulting in an increase in pore pressure.

Generally, the rates of stress increase and pore water expulsion are consistent, up to a maximum determined by the permeability of the rock at any particular time. If this hydraulic conductivity is exceeded by pore volume reduction, the magnitude of the resultant geopressure is a direct function of their differences.

The manner in which a sedimentary section deforms under tectonic stress will produce different pore pressure regimes. Figure 2-14 shows geopressure distribution within a series of active compressional folds. If thrust or reverse faulting is developed in preference to folding, then geopressures may develop in clay strata anywhere in the tectonic region, particularly in zones where for some reason permeability is relatively restricted. This figure indicates that geopressures occur initially within the hinge portion of each compressional fold in a thick clay sequence. Initially, bedding planes may act as slip planes, and the spacing of these slip planes will determine the amount of physical shortening or extension, below and above a slip plane, respectively. Hence, if the clay strata are massive the slip planes will be few, causing more shortening in the hinge zones than there would be if the strata were thinly bedded. Active compressional folds of this type occur offshore Baluchistan in the Arabian Sea; they are adjacent to a major ocean floor transcurrent fault, and the movement along the fault has provided the energy for the folding.

In many cases faults act as hydraulic and hydrocarbon conduits, while in other cases they may act as seals. Hence a normally faulted basin may or may not be geopressured, and may or may not contain hydrocarbons. Normal faulting occurs in response to a relaxed tectonic environment and may form simply due to differential compaction. However, a very slight tensional tectonic stress (as would exist over high points in buried topography, domes, domal uplift) also causes normal faulting and, in the extreme, graben structures. In these instances the fault surfaces may be relaxed (not held together by the weight of the overburden), in which case they cannot act as a seal to a reservoir or geopressure. If this type of faulting does occur after the formation of a hydrocarbon reservoir or geopressure, then the oil and excess pressure will be released rapidly because the permeability along fracture/fault conduits is extremely large.

Tectonically produced geopressures will behave and appear much the same as those resulting from (vertical) subcompaction. Clues to their origin may be found in signs of deformation in regional structure and rock fabric. The presence of a regional tectonic stress will also be indicated by the results of formation fracture tests (see Section 5).

## 2.9 PIEZOMETRIC CHANGES

A piezometric or potentiometric surface is an imaginary surface that represents the static head of groundwater and is defined by the level to which water will rise in a well. The water table is a particular potentiometric surface. However, the water table represents the potentiometric surface for the adjacent formations and their relation to the topography, but this does not represent the static head that would be caused by geopressured strata.

Groundwaters are dynamic, and flow occurs from high to low hydraulic potential. This dynamic condition is created by changes in topography, climatic fluctuations, sea level changes, isostasy and tectonics. Flowing groundwater creates a pressure which is additional to hydrostatic. For most purposes in geopressure evaluation it must be assumed that the hydrostatic pressure gradient is that of a static column, as there is no way to measure the pressure differences caused by dynamic groundwater. However, the actual flowpath of groundwaters is tortuous, hence lines joining points of equal pressure will not be horizontal as assumed in geopressure evaluation. Actual hydrostatic pressure will increase with increasing tortuosity of these isobars and will include the overall immeasurable frictional pressure loss caused by flowing pore waters.

The normal hydrostatic pressure gradient is thus a function of the density of the pore fluid, its velocity, and the tortuosity of flowpaths. Pore fluid movement is primarily governed by pressure potentials, but electro-osmotic relationships and grain character also play a major part. Changes in fluid velocity may be caused by the physical phenomena mentioned above, resulting in changes in the pore fluid isobars which, in turn, are reflected by a change in the potentiometric surface.

In summary, changes in the potentiometric surface for a particular zone affect the normal hydrostatic gradient; its most noticeable effect is onshore in arid areas and also in near-shore environments that will be affected by tides, river volume fluctuations, and climatic variations. Figure 2-15 illustrates the overlapping potentiometric surfaces with a "normally" pressured and geopressured area.

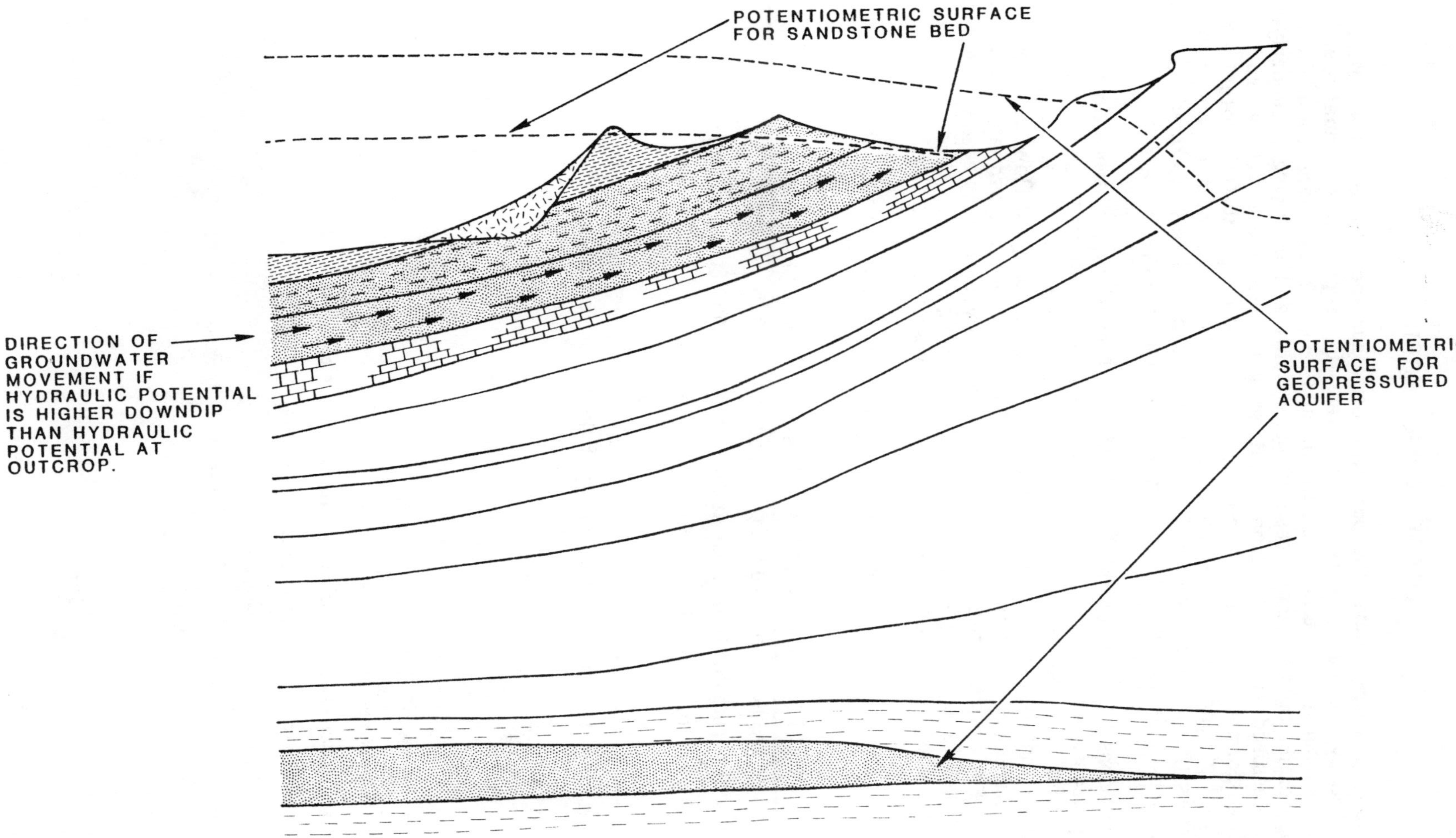


Figure 2-15. Overlapping potentiometric surfaces with normally pressured and geopressured zones

## 2.10 THE TRANSITION ZONE

In a normally pressured zone, the potentiometric surface is constant and may be approximated by the water table or sea level. In geopressured formations, the potentiometric surface is higher than the normal level. Thus, there is a difference in fluid potential between geopressured pore fluids and overlying or underlying normally pressured fluids. This potential gradient will cause a flow from the geopressured formation to the lower-pressured formations, with fluid potentials in the intervening zone varying between normal and the level of the geopressured zone. Unless there is a perfect seal of zero permeability, this intermediate zone will be present. It is known as the pressure transition zone.

Since formations of zero permeability rarely (if ever) occur, transitions zones commonly occur. The thickness of this zone will depend upon the permeabilities (1) within and adjacent to the geopressured formation and (2) the age of the geopressure; that is, the time that has been available for fluid flow and pressure depletion. Remember that the age of the geopressure may be less than that of the formation. For example, tectonism or epierogenic movement may result in lateral stress, changes in depth of burial, or increases in temperature. All of these may result in the geopressuring of a previously normal pressured formation.

To illustrate the dynamic nature of the transition zone, consider the following example:

Pressure at 3000 ft is normal, 8.5 lb/gal EMD or 1323 psi

Pressure gradient from 4000 ft to 5000 ft is 14.0 lb/gal EMD

Vertical permeability in shale from 3000 ft to 4000 ft is $10^{-6}$ md

Mean porosity of the geopressured zone is 25%

Darcy's Law gives the rate of dewatering of the geopressured zone:

$$Q = K \frac{A}{\mu L} * \Delta P \tag{2-2}$$

where

$Q$ = rate of fluid flow ($cm^3$/sec)

$K$ = permeability (darcies)

$A$ = cross-sectional area (assume 1 $cm^2$)

$\mu$ = fluid viscosity (assume 1 cps)

$L$ = length of flowpath (cm)

$\Delta P$ = difference in fluid potential (atmospheres)

In the above example, $L = (4000-3000) * 30.48 = 30{,}480$ cm

$$P = (14.0-8.5) * 4000 * .0519 \div 14.7 = 77.7 \text{ atm}$$

$$\therefore \quad Q = 10^{-9} * \frac{77.7}{30{,}480} = 2.55 * 10^{-12} \text{ cm}^3/\text{sec}$$

Total fluid volume in the geopressured reservoir is $1000 * .25 * 30.48$

$$= 7620 \text{ cm}^3, \text{ per cm}^2 \text{ area}$$

$$\therefore \quad \text{Rate of dewatering} = \frac{2.55 * 10^{-12}}{7620} * 100 = 3.35 * 10^{-14}\% \text{ per second}$$

Figure 2-7 indicates that a 1% change in specific volume is equivalent to a relaxation of 3000 psi pressure, so in this example the rate of pressure relief may be calculated as

$$\frac{1}{3.35} * 10^{14}/3000 \text{ seconds per 1 psi,}$$

or 316 years per 1 psi (66,000 years per 1 lb/gal EMD).

Such a rate is well within the possibilities of geopressure generation, thus the existence of a transition zone need not significantly affect development of high pore pressure gradients.

The presence of a transition zone is of great importance to the pressure evaluation geologist. It is a zone in which rock properties may exhibit a gradual change from normally pressured trends to values typical of geopressures. In many cases, and particularly when compaction proceeds simultaneously with geopressure generation, the transition zone will take on all the characteristics of a geopressured formation. This will usually be the case when geopressure is generated primarily by compaction disequilibrium. In other cases, such as tectonic pressuring, the transition zone may not show the relatively high porosity characteristic of geopressures; nevertheless, its drilling response will permit use of most of the techniques described in Section 4.

It is important to be aware that, in terms of pressure evaluation, the transition zone is <u>geopressured</u>. Although it may have a secondary origin and may have physical and even lithological differences from the original sealed zone, it nevertheless <u>now</u> has a pore pressure in excess of normal hydrostatic. In the past, the term "geopressure prediction" was used to describe the identification of a transition zone as if it were an advance warning of a geopressure rather than the beginning of it. Logging personnel should avoid using the word "prediction" and point out the error to anyone who uses it to describe our services.

## 2.11 PORE PRESSURE MAINTENANCE

Paragraphs 2.2 through 2.9 outlined the major processes which may act uniquely or in combination to form geopressures of any magnitude. In ideal situations, the various processes may cause local pressure gradient increases to

- overburden pressure gradient; specifically, compaction disequilibrium
- twice the overburden pressure gradient by means of aquathermal pressuring and possibly montmorillonite dehydration
- any gradient caused by tectonics, osmosis, various combinations of the processes

Exploration has shown that geopressures are mainly confined to young rocks, suggesting that these instabilities are geologically transient affairs which, once formed, do not last for eternity. The questions then arise: Is the particular geopressure encountered now in the process of formation? At the peak of its generation? Has pressure dissipation through leakage occurred, lowering the geopressure gradient?

Geopressures formed due to compaction disequilibrium and aquathermal pressuring theoretically can continue to increase with increasing burial and temperature. Pressure gradients produced by compaction disequilbrium parallel the overburden pressure gradient only if no leakage has occurred. This is the ideal case, and the effective overburden pressure (see paragraph 3.6) remains constant with burial. Slight leakage (normally the case) results in pressure gradients less than the overburden pressure gradient, signifying that the effective overburden pressure is increasing. However, geopressures formed due only to aquathermal pressuring cannot increase beyond the overburden pressure, as any subsequent increase will cause horizontal fracturing and leakage of the excess pore fluid.

The water released from montmorillonite as it undergoes diagenesis to illite is a finite amount; hence in an isolated environment the rates will cause a finite pressure increase. This is a "one shot" event. Overall, it may be a series of water effluxes — but once completed, no more water will be produced.

Geopressures caused by tectonics, osmosis, or various combinations may form any magnitude of pore pressure below that of the overburden. It is thus important to distinguish between steady state and transient conditions. Some geopressures obviously prevail over long geological time intervals, whereas others are only temporary phenomena and in some cases may be geologically instantaneous. Conditions leading to very short-lived geopressures are rapid fault movement, loading by landsliding, loading by turbidities, freezing of talik, etc. (Gretener, 1977). Rapid pore pressure fluctuations may leave a permanent record:

- conditions created that are suitable for ductile deformation, flow, or both
- disturbance of recently deposited sediment
- modification of fault geometry

It is probably reasonable to say that geopressures occurring in shallow sediments offshore are still in their infant stages. Abnormal pore pressure zones that occur in consolidated rock in a subsiding basin may be either developing, mature, or dissipating. If dissipation is occurring, zones of sufficient hydraulic conductivity must have been created and must be at a lower hydraulic potential, or the rate of pore pressure increase is less than the overall hydraulic conductivity of the geopressured rock.

It requires extremely detailed analysis to conclusively identify the particular stage at which a geopressured zone exists. However, the Pressure Evaluation Geologist should keep in mind that certain processes continually promote the generation of abnormal pore pressures, whereas others can occur only over a finite period.

A geopressured zone contains potential energy at a level higher than adjacent, normally pressured rocks; and to increase this imbalance, more energy must be supplied to the zone. This energy imbalance is continually seeking equalization, such that overall equilibrium may be reobtained. Further, the occurrence of geopressures should not be considered abnormal, according to Price (1978). Price concludes that, in any sedimentary sequence containing appreciable quantities of muds, clays, evaporites or other relatively impermeable material, high geopressures are a natural consequence of dewatering. Hence these high pressures, although being transient, are certainly not anomalous because they will be a feature of most sedimentary sequences during compaction.

The processes that can maintain an energy imbalance must, in time, become weaker, and it has been shown that through the geopressure generating processes themselves, limiting factors are inherent.

## 2.12 REFERENCES

Barker, C., 1972, Aquathermal Pressure — Role of Temperature in Development of Abnormal-Pressure Zones, AAPG Bull., v. 56, n. 10.

Bradley, J. S., 1975, Abnormal Formation Pressure, AAPG Bull., v. 59, n. 6.

Brederhoeft, J. D. et al, 1963, Possible Mechanisms for Concentration of Brines in Subsurface Formations, AAPG Bull., v. 47, n. 2.

Burst, J. F., 1969, Diagenesis of Gulf Coast Clayey Sediments and Its Possible Relation to Petroleum Migration, AAPG Bull., v. 53, n. 1.

Deer, W. A. et al, 1967, An Introduction to the Rock Forming Minerals, Longmans Press.

Fertl, W. H., 1973, Abnormal Formation Pressures, Elsevier Press.

Fyfe, W. S., N. J. Price and A. B. Thompson, 1978, Fluids in the Earth's Crust, Elsevier Scientific Pub. Co.

Gretener, P. E., 1978, Pore Pressure: Fundamentals, General Ramifications and Implications for Structural Geology, AAPG Course Note Series 4.

Hanshaw, B. B., and E. Zen, 1965, Osmotic Equilibrium and Overthrust Faulting, Geol. Soc. Am. Bull., v. 76, n. 12.

Hinch, H. H., 1978, The Nature of Shales and the Dynamics of Hydrocarbon Expulsion in the Gulf Coast Tertiary Section, AAPG Course Note Series 8.

Hubbert, M. K., 1972, Structural Geology, Hafner Pub. Co.

Jones, P. H., 1969, Hydrodynamics of Geopressures in the Northern Gulf of Mexico Basin, J. P. T., July.

Kennedy, G. C., and W. T. Holser, 1966, Pressure-Volume-Temperature and Phase Relations of Water and Carbon Dioxide, Geol. Soc. Am. Memoir 97.

Lewis, C. R., and S. C. Rose, 1970, A Theory Relating High Temperatures and Overpressures, J. P. T., January.

Magara, K., 1975(b), Reevaluation of Montmorillonite Dehydration as Cause of Abnormal Pressure and Hydrocarbon Migration, AAPG Bull., v. 59, n. 2.

Magara, K., 1975(a), Importance of Aquathermal Pressuring Effect in Gulf Coast, AAPG Bull., v. 59, n. 10.

McCain, W. D., 1973, The Properties of Petroleum Fluids, The Petroleum Publishing Company.

Olsen, H. W., 1972, Liquid Movement Through Kaolinite under Hydraulic, Electric and Osmotic Gradients, AAPG Bull., v. 56, n. 10.

Powers, M. C., 1967, Fluid-Release Mechanisms in Compacting Marine Mudrocks and their Importance in Oil Exploration, AAPG Bull., v. 51, n. 7.

Rieke III, H. H., and G.V. Chilingarian, 1974, Compaction of Argillaceous Sediments, Developments in Sedimentology, 16, Elsevier Press, New York.

Rubey, W. W., and M. K. Hubbert, 1959, Overthrust Belt in Geosynclinal Area of Western Wyoming in Light of Fluid Pressure Hypothesis, Geol. Soc. Am. Bull., v. 70.

Weaver, C. E., and K. C. Beck, 1971, Clay Water Diagenesis During Burial: How Mud Becomes Gneiss, Geol. Soc. Am. Special Paper 134.

Young, A., and P. F. Low, 1965, Osmosis in Argillaceous Rocks, AAPG Bull., v. 49, n. 7.

# 3 ENGINEERING

## 3.1 INTRODUCTION

The preceding section dealt (somewhat academically) with the major accepted theories that provide explanations for some of the various anomalous phenomena that may be encountered in oil exploration. Some of these theories originated in the laboratory, and some were developed from field experience. In order that these ideas may be gainfully employed by the geologist for geopressure evaluation, it is necessary to provide numerical expressions that enable the academic theories to be workable in an engineering environment. The task at hand is to apply geological training to an engineering field, and in doing so the geologist must work closely with the client engineers. Communication with engineers and geologists requires various degrees of tact and diplomacy at all times. The P.E.G. and GEMDAS operator must endeavor to maintain efficient communication channels at all times, for without this communication, the value of Exlog's service may become severely eclipsed.

The Pressure Evaluation Geologist (P.E.G.) has to communicate effectively with an experienced engineer in order that any ambiguity can be avoided. To facilitate this, it is the geologist's responsibility (as outlined in Section 1) to observe, learn, and then act. It is also the geologist's responsibility to know the rig activity at all times. All of this involves communication, and it is vital to the success of the operation.

The following paragraphs describe subsurface phenomena in mathematical notation, and the resulting formulae form the backbone to the P.E. geologist's calculations.

## 3.2 SUBSURFACE PRESSURES

### 3.3 HYDROSTATIC PRESSURE

Hydrostatic pressure (or hydropressure) is defined as the pressure exerted by the water at any given point in a body of water at rest. <u>It is the pressure due to the density and vertical height of the fluid column</u> (see Figure 3-1).

$$P = 0.0519 * W * D \qquad (3\text{-}1)$$

where

P = hydrostatic pressure (psi)

W = water (or fluid) density (lb/gal)

D = vertical depth (ft)

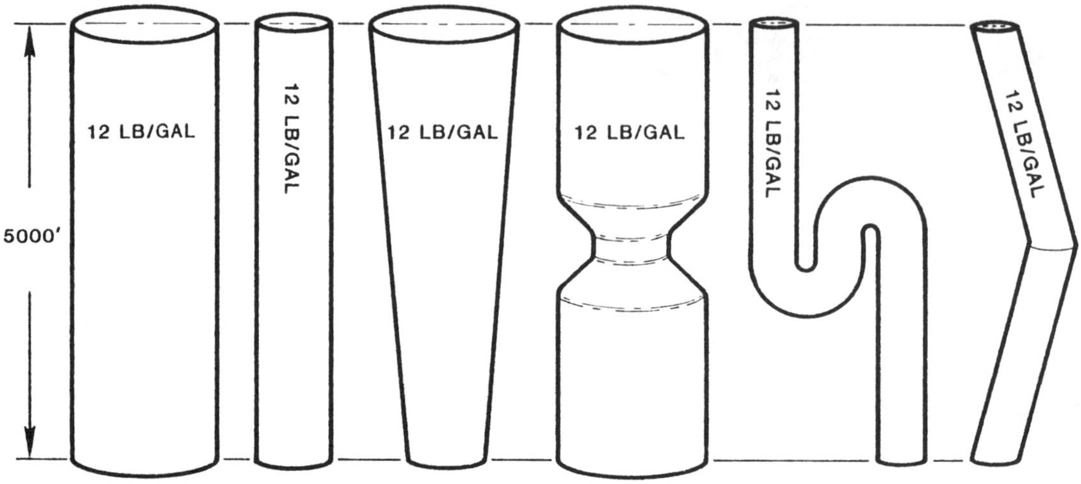


Figure 3-1. Hydrostatic pressure (P) is a function of the density and vertical height of the fluid column

The number 0.0519 is a conversion factor for oilfield imperial units (psi, lb/gal, ft) and is derived as follows:

There are 7.48 gallons in 1 cu ft
There are 144 sq inches in 1 sq ft

hence

$$\text{lb/gal} * 7.48\ \text{gal/ft}^3 * \frac{1}{144}\ \text{ft}^2/\text{in.}^2 = \text{psi/ft} \tag{3-2}$$

therefore

$$\frac{7.48}{144} = \text{psi/ft/lb/gal} \tag{3-3}$$

$$0.0519 = \text{psi/ft/lb/gal} \tag{3-4}$$

So fresh water, having a density of 8.34 lb/gal, or 62.35 lb/cu ft, exerts a pressure of

$$8.34 * 0.0519 = 0.433\ \text{psi/ft} \tag{3-5}$$

Similarly, using S.I. units:

$$P(\text{kPa}) = W(\text{kg/m}^3) * D(\text{m}) * 0.0098 \tag{3-6}$$

It is important to realize that the units g/cc, lb/gal, psi/ft, or their equivalents express a gradient or pressure per unit depth. However, it is common to refer to the mud density as mud weight, still being expressed in lb/gal or g/cc. Note that specific gravity (SG) is not a density. SG is the ratio of a density over the density of water, and hence has no units. Oilfield accuracy tolerances, however, allow SG to be numerically equal to the material's density in g/cc. It is convenient to relate various pressures by the resultant gradient they produce relative to a fixed datum, usually RKB (rotary kelly bushing), i.e. from the rig floor. Since the mud level (the flowline) is below RKB, all gradients are referenced from the flowline. Usually the distance from RKB to flowline is around 5 to 10 feet, but this is sufficient to cause gradient differences at shallow depths. At greater depths, the distance from RKB to flowline becomes insignificant when calculating gradients, but, for sake of consistency, all gradients are calculated from the flowline. Also, the formation balance gradient is usually spoken of in pounds per gallon; this makes comparisons with mud density simple. However, when writing reports, take care to use the correct terminology — formation balance gradient at 10,000 ft is 15.7 lb/gal EMD, pore pressure at 10,000 ft is 8148 psi.

The gradient that the pore fluid density produces alone is called normal pore pressure gradient. Hence, this gradient is dependent upon the density of the pore waters and will vary from area to area.

Onshore (e.g., Rocky Mountain area), the water is relatively fresh, hence the normal pore pressure gradient is approximately

$$8.34 * 0.0519 = 0.433 \text{ psi/ft}$$

In the Gulf Coast area, waters are more saline and the normal pore pressure gradient is

$$8.96 * 0.0519 = 0.465 \text{ psi/ft}$$

In other offshore areas, seawater density and pore water density may vary from slightly saline (8.5 lb/gal) to saturated saline (9.9 lb/gal). Since salinity varies with depth and formation, the average value may not be valid for all depths; and in each area a log-data-derived pressure-versus-depth profile must be found, (see paragraph 3.9).

Salinity of formation water may also be dependent upon lithology. In certain evaporites, saturated saltwater has a gradient of 0.520 psi/ft. Therefore, knowledge of the environment is important. For example, in a Zechstein basin a calculated pressure gradient of 0.520 psi/ft would not be very significant, whereas in a fresh-water basin it would indicate a large overpressure (Figure 3-2).

Before a well is drilled, an estimate of the normal pore pressure gradient must be found. This may be obtained from actual density measurements, direct pressure measurements from offset wells, SP and resistivity log interpretation, or by assuming that the density is the same as seawater if offshore, or that onshore the water is fresh. If the well is a rank wildcat and no previous data is available, assume that the normal pore pressure gradient is 8.34 lb/gal if onshore, or seawater density if offshore (8.5 to 9 lb/gal).

| Formation Water Type | Salinity Chloride mg/litre | ppm NaC1 | Normal Pressure Gradient (psi/ft) | Equivalent Mudweight (lb/gal) |
|---|---|---|---|---|
| Fresh Water | 0 | 0 | 0.433 | 8.34 |
| Brackish Water | 6,098<br>12,287<br>24,921 | 10,062<br>20,273<br>41,120 | 0.435<br>0.438<br>0.444 | 8.37<br>8.43<br>8.55 |
| Seawater | 33,000 | 54,450 | 0.448 | 8.63 |
| Saltwater | 37,912<br>51,296<br>64,987 | 62,554<br>84,638<br>107,228 | 0.451<br>0.457<br>0.464 | 8.67<br>8.80<br>8.92 |
| Typical Offshore Gradient | 65,287<br>79,065<br>93,507<br>108,373<br>123,604<br>139,320<br>155,440<br>171,905<br>188,895 | 107,709<br>130,457<br>154,286<br>178,815<br>203,946<br>229,878<br>256,476<br>283,643<br>311,676 | 0.465<br>0.470<br>0.477<br>0.484<br>0.490<br>0.497<br>0.504<br>0.511<br>0.518 | 8.96<br>9.04<br>9.17<br>9.30<br>9.43<br>9.56<br>9.71<br>9.83<br>9.97 |
| Saturated Saltwater | 191,600 | 316,140 | 0.519 | 9.99 |

Figure 3-2. Variation of hydrostatic pressure with formation water salinity

For example, if the normal pore pressure gradient is 8.34 lb/gal, then the pore pressure at 5000 feet is

$$5000 * 8.34 * 0.0519 = 2164 \text{ psi}$$

If the normal pore pressure gradient is 8.7 lb/gal, the pore pressure at 5000 feet is

$$5000 * 8.7 * 0.0519 = 2258 \text{ psi}$$

Note that the apparent small change in pore pressure gradient produces a large change in pore pressure at depth. This is accentuated as depth increases, hence it is very important that accurate normal pore pressure gradients are obtained.

### 3.4 OVERBURDEN PRESSURE

The overburden pressure at any point in the formation is that pressure exerted by the total weight of the overlying formations. This is expressed as:

$$S = 0.433 \int_0^z \rho(z)\, dz \qquad (3\text{-}7)$$

where

$S$ = overburden pressure (psi)

$z$ = depth interval (ft)

$\rho$ = bulk density (g/cc)

The bulk density of a rock is a function of the density of the rock matrix itself, the density of the pore fluids, and the porosity.

$$\rho_b = \emptyset \rho_f + (1-\emptyset)\, \rho_m \qquad (3\text{-}8)$$

where

$\rho_b$ = bulk density (g/cc)

$\emptyset$ = porosity (fractional)

$\rho_m$ = density of the matrix (g/cc)

$\rho_f$ = density of the pore fluid (g/cc)

From this relationship it can be seen that as $\emptyset$ approaches 1, $\rho_b$ approaches $\rho_f$; and conversely, as $\emptyset$ approaches 0, $\rho_b$ approaches $\rho_m$. If all the densities are known, the porosity may be found:

$$\emptyset = \frac{\rho_m - \rho_b}{\rho_m - \rho_f} \qquad (3\text{-}9)$$

Typical densities of rocks and fluids are listed in Figure 3-3.

| Lithology | Matrix Density |
|---|---|
| Sandstone | 2.65 |
| Limestone | 2.71 |
| Dolomite | 2.87 |
| Anhydrite | 2.98 |
| Halite | 2.03 |
| Gypsum | 2.35 |
| Clay | ≈2.7 - 2.8 |
| Fresh Water | 1.0 |
| Salt Water | 1.15 |
| Oil | 0.80 |

Figure 3-3. Typical densities of rocks and fluids

Overburden pressures are again measured from the rig floor, hence offshore the depth and density of seawater must be taken into account.

The overburden pressure gradient is the average density to the depth of interest. Because of compaction, the overburden gradient is not constant. However, as $\rho_b$ approaches $\rho_m$ due to very low porosity, compaction with further depth will be exceedingly slow; hence the overburden pressure gradient will become asymptotic to $\rho_m$. Since the average density of a thick sedimentary sequence is approximately 2.3 g/cc, with depth the overburden gradient will be about 19.2 lb/gal or 1 psi/ft:

$$2.3 \text{ g/cc} * 8.34 = 19.2 \text{ lb/gal}$$

or

$$2.3 * 0.433 = 1 \text{ psi/ft}$$

Onshore, with more compact sediments, the overburden pressure gradient may be assumed to be close to 1 psi/ft, but offshore the actual overburden gradient will be much less than this due to the effect of seawater, the air gap, and large thicknesses of unconsolidated sediment. Depending on the water depth, offshore overburden pressure gradients may be as low as 14 lb/gal or 0.73 psi/ft at 5000 feet.

The overburden pressure may be calculated as follows:

$$S = 0.433 * \rho_b \text{ average} * D \tag{3-10}$$

where

S = overburden pressure for interval, D (psi)

D = depth interval (ft; usually 100 ft is sufficient but varies according to lithological changes)

$\rho_b$ = average bulk density for interval (g/cc)

0.433 = constant for converting g/cc to psi/ft.

This pressure is then converted to a gradient through the relationship

$$\text{OBG} = \frac{\sum S}{\sum D} = \frac{\sum 0.433 * \rho_b \text{ average} * D}{\sum D} \tag{3-11}$$

Equation (3-11) describes an estimate of the average overburden pressure gradient.

For example: A rig is offshore in 470 feet of water. The distance from the flowline to sea level is 65 feet.

(a) seawater depth = 470 ft
seawater density = 1.03 g/cc

From Equation (3-10),

$S = 0.433 * 1.03 * 470$
$S = 210$ psi

then, with Equation (3-11),

$$\text{OBG} = \frac{210}{470 + 65}$$

OBG = 0.393 psi/ft or 7.6 lb/gal

Note the effect that the air gap has on the calculated gradient. The seawater density is 1.03 * 8.34 = 8.6 lb/gal, but, because of the 65-ft air gap, the overall gradient from the rig floor is only 7.6 lb/gal, to the seabed.

(b) first formation increment = 200 ft
average formation density = 1.9 g/cc

From Equation (3-10),

$$S = 0.433 * 1.9 * 200$$
$$S = 165 \text{ psi}$$

From Equation (3-11),

$$OBG = \frac{165 + 210}{470 + 65 + 200}$$

$$OBG = 0.510 \text{ psi/ft or } 9.8 \text{ lb/gal}$$

(c) second formation increment = 200 ft
average formation density = 1.95 g/cc

From Equation (3-10),

$$S = 0.433 * 1.95 * 200$$
$$S = 169 \text{ psi}$$

From Equation (3-11),

$$OBG = \frac{169 + 210 + 165}{470 + 65 + 400}$$

$$OBG = 0.581 \text{ psi/ft or } 11.2 \text{ lb/gal}$$

In ELOS,® an EAP program is available to rapidly perform these repetetive calculations (see Figure 3-4).

Accurate bulk density values must be obtained in order that the overburden gradient is as accurate as possible. If a density log is run, bulk densities may be averaged easily and read off for each interval. If a density log is not available, bulk densities may be calculated from the sonic log via porosity. Refer to the discussion on wireline logging in Section 4 for this method. As a last resort, actual bulk densities from cuttings may be used; however, this method is normally inaccurate due to possible hydration of clay cuttings in the annulus and consequent swelling, causing erroneously low bulk density values.

Since all rocks contribute to the overburden pressure, bulk density readings should not be limited to clay beds. Take values from all the lithologies present in the interval, and average them.

IRWELL VALLEY DRILLING CONSORTIUM : TEST WELL # 1
3:49 5/ 9/81

OVERBURDEN GRADIENT CALCULATION

RKB TO MEAN SEA LEVEL = 55.0ft
RKB TO SEA BED = 500.0ft
SEA WATER DENSITY = 1.10g/cc

| DEPTH | | INTERVAL | DENSITY | OVERBURDEN | | |
|---|---|---|---|---|---|---|
| FROM(ft) | TO(ft) | ft | g/cc | psi | psi/ft | lb/gal |
| 500.0 | 550.0 | 50.0 | 2.22 | 260 | .473 | 9.12 |
| 550.0 | 600.0 | 50.0 | 2.24 | 309 | .515 | 9.92 |
| 600.0 | 650.0 | 50.0 | 2.23 | 357 | .550 | 10.59 |
| 650.0 | 700.0 | 50.0 | 2.25 | 406 | .580 | 11.18 |
| 700.0 | 750.0 | 50.0 | 2.24 | 455 | .606 | 11.68 |
| 750.0 | 800.0 | 50.0 | 2.24 | 503 | .629 | 12.12 |
| 800.0 | 850.0 | 50.0 | 2.25 | 552 | .649 | 12.51 |
| 850.0 | 900.0 | 50.0 | 2.29 | 602 | .668 | 12.88 |
| 900.0 | 950.0 | 50.0 | 2.31 | 652 | .686 | 13.22 |
| 950.0 | 1000.0 | 50.0 | 2.25 | 700 | .700 | 13.49 |
| 1000.0 | 1050.0 | 50.0 | 2.27 | 750 | .714 | 13.75 |
| 1050.0 | 1100.0 | 50.0 | 2.29 | 799 | .727 | 14.00 |
| 1100.0 | 1150.0 | 50.0 | 2.29 | 849 | .738 | 14.22 |
| 1150.0 | 1200.0 | 50.0 | 2.28 | 898 | .749 | 14.42 |
| 1200.0 | 1250.0 | 50.0 | 2.27 | 947 | .758 | 14.60 |
| 1250.0 | 1300.0 | 50.0 | 2.28 | 997 | .767 | 14.78 |
| 1300.0 | 1350.0 | 50.0 | 2.26 | 1046 | .775 | 14.93 |
| 1350.0 | 1400.0 | 50.0 | 2.27 | 1095 | .782 | 15.07 |
| 1400.0 | 1450.0 | 50.0 | 2.27 | 1144 | .789 | 15.21 |
| 1450.0 | 1500.0 | 50.0 | 2.29 | 1194 | .796 | 15.34 |
| 1500.0 | 1550.0 | 50.0 | 2.32 | 1244 | .803 | 15.47 |
| 1550.0 | 1600.0 | 50.0 | 2.31 | 1294 | .809 | 15.59 |
| 1600.0 | 1650.0 | 50.0 | 2.31 | 1344 | .815 | 15.70 |
| 1650.0 | 1700.0 | 50.0 | 2.28 | 1394 | .820 | 15.80 |
| 1700.0 | 1750.0 | 50.0 | 2.29 | 1443 | .825 | 15.89 |
| 1750.0 | 1800.0 | 50.0 | 2.29 | 1493 | .829 | 15.98 |
| 1800.0 | 1850.0 | 50.0 | 2.26 | 1542 | .834 | 16.06 |
| 1850.0 | 1900.0 | 50.0 | 2.30 | 1592 | .838 | 16.14 |
| 1900.0 | 1950.0 | 50.0 | 2.30 | 1642 | .842 | 16.22 |

O.B.G. DEPTH EQUATION: LN(DEPTH) = LN(A) + B * (O.B.G.)
WHERE A = 8.940495E+01
B = 3.544996E+00
COEFFICIENT OF DETERMINATION = .98

Figure 3-4. Typical overburden pressure gradients

Overburden pressure at any point may be calculated from the overburden pressure gradient at that point. For example:

```
At 10,000 feet, OBG   = 17.5 lb/gal or 0.908 psi/ft
  Overburden pressure = 17.5 * 10,000 * 0.0519 = 9083 psi
                      = 0.908 * 10,000 = 9080 psi
```

This example also serves to illustrate rounding errors in the calculations. There is no point in calculating figures to an accuracy comparable to that used in atomic physics when the accuracy of rig instruments may be as low as 90 percent and, at best, 95 percent. Thus it is necessary to calculate pressures only to the nearest psi, or gradients to the nearest 10th (or possibly 100th) of a lb/gal.

An overburden pressure gradient curve (in either lb/gal or psi/ft) will be similar to those shown in Figure 3-5. Variations result due to overall bulk density (source rock type and compaction), water depth, rig elevation, pore pressures, and tectonic regime. These curves are shown for illustrative and comparative purposes only, and should not be used as a reference for a particular overburden gradient curve if an FDC log is not run. As shown in the diagram, considerable variations are apparent — largely caused by different water depths and different air gaps; hence all effort must be made either to obtain an overburden gradient curve from a nearby well drilled with the same rig, or to develop an overburden gradient from the sonic log (see Section 4, paragraph 4.16).

Figure 3-6 illustrates these same overburdens, but plotted as overburden pressure against depth. Note that all the trends are approximately parallel except for the upper 4000 feet. Also, below this depth, the pressure/depth curves are very close to linear. Very slight deviations from a linear curve produce variable overburden gradients (or vice versa). Transposing an overburden gradient plot from an overburden pressure plot is difficult due to the scales involved; however, an overburden pressure plot may be linearly extrapolated to the next logging point with a reasonable degree of accuracy, for the purpose of kick-tolerance/pore pressure/fracture-pressure planning (discussed in Section 5).

Overburden gradient reversals may occur over short intervals in thick salt sequences and in very high pore pressure zones. Geopressures usually occur in claystone, and, due to the abnormally high porosity, the bulk density is exceedingly low. If the zone is thick enough, the amount of low density readings may be sufficient to cause the overall average bulk density (literally, the overburden pressure gradient curve) to decrease. Since these low density zones are usually relatively thin, the overburden gradient reversal is normally small and occurs over a short interval. Note that it is the gradient that decreases: the overburden pressure still increases over the zone, but the rate of increase is slower than that above.

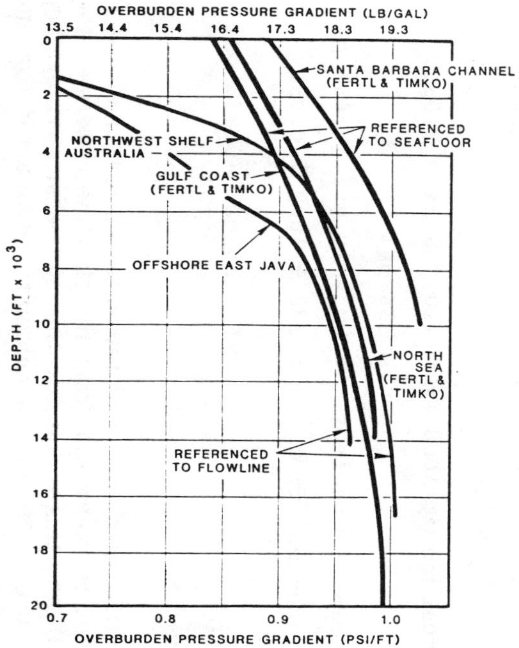


Figure 3-5. ELOS Overburden Gradient Calculation

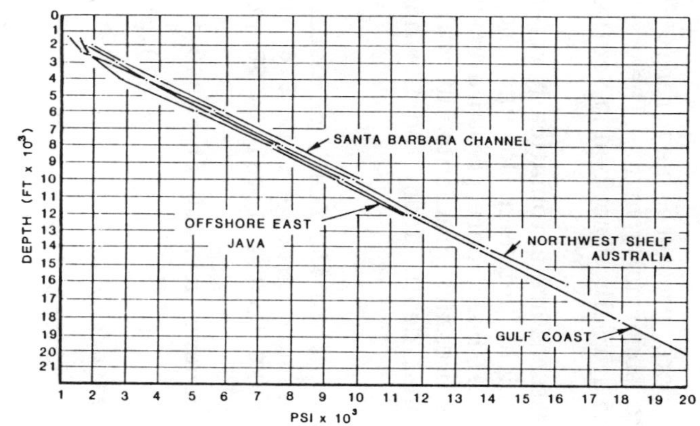


Figure 3-6. Relationship between normal pore pressure gradient, normal formation balance gradient, formation balance gradient, and equivalent mud density

## 3.5 PORE PRESSURES, FORMATION BALANCE GRADIENT, AND EQUIVALENT MUD DENSITY

As previously stated, pore pressure may range from abnormally low, through hydrostatic (normal) to abnormally high. The various evaluation techniques for abnormal pore pressures are fully described in Section 4.

While drilling, pore pressures are automatically referenced to flowline. Hence, due to the differences in height between flowline and the water table onshore, and flowline and sea level offshore, pore pressure gradients measured during drilling will not be actual pore pressure gradients but will represent that pressure that a fluid will balance to that depth, from the flowline. This can be illustrated as follows:

- Offshore: water depth 450 feet, seawater density 8.6 lb/gal. Assume normal pore pressure gradient is 8.6 lb/gal, depth of interest is 1000 ft, RKB. RKB to sea level is 60 ft, RKB to flowline is 5 ft.

  actual pore pressure at 1000 ft = 940 * 8.6 * 0.0519
  = 420 psi

  actual pore pressure gradient = 8.6 lb/gal (0.446 psi/ft)

  pore pressure gradient from flowline = 420 ÷ 995 ÷ 0.0519
  = 8.1 lb/gal (0.422 psi/ft)

- Onshore: Depth to water table is 220 feet, water density is 8.34 lb/gal, flowline to ground level is 45 feet, depth of interest is 1000 feet (below flowline).

  actual pore pressure at 1000 ft = 780 * 8.34 * 0.0519
  = 338 psi

  actual pore pressure gradient = 8.34 lb/gal (0.433 psi/ft)

  pore pressure gradient from flowline = 338 ÷ 1000 ÷ 0.0519
  = 6.5 lb/gal (0.338 psi/ft)

It is clear that, at shallow depths, the differences are extremely important. For this reason, the gradient as measured from the flowline is termed the Formation Balance Gradient (FBG), and this is equal to the mud density required in the hole to balance the pore pressure. The values calculated above for hypothetical cases (both on- and offshore), 6.5 lb/gal and 8.1 lb/gal, are thus the mud densities needed to balance the pore pressures at 1000 feet for those conditions. Obviously, no water-based drilling mud can be as light as these, and this represents a major problem in drilling shallow holes in that fracture pressures are often approached and exceeded, resulting in no returns. (Section 5 offers a complete explanation.)

The Formation Balance Gradient is the pore pressure gradient referenced to the flowline, and when expressed in terms of mud density (lb/gal), it expresses that mud density which is necessary to balance the pore pressure at the depth of interest. Figure 3-7 shows an example worksheet for calculating normal FBG. Figure 3-8 shows the relation between the actual fluid density and the FBG (EMD), the gradient referenced to the flowline.

Some of these terms may be used synonymously, which results in confusion if they are not fully understood:

- "Local" Pressure Gradient is used in this manual to describe the actual rate of pressure change with depth at a point in the formation. Where fluid communication exists it is simply the hydrostatic pressure gradient; when expressed in units of density it is equal to the actual fluid density present at the point. Only when pressure is increasing at a non-hydrostatic gradient (that is, in a transition zone) will it be higher. Although the only "true" gradient (that is, rate of change) term, it is rarely directly applicable to wellsite pressure calculations. The following gross gradient (that is, pressure divided by depth) terms have more common practical use.

- Pore Pressure Gradient is pressure per unit depth, measured from the top of the formation fluid column. Onshore it is measured from the level of the water table, and offshore it is measured from the sea level.

- Formation Balance Gradient is pressure per unit depth, measured from the flowline. It is precisely equal to Equivalent Mud Density (EMD), so the terms may be used interchangeably. The Formation Balance Gradient is thus always less than the Pore Pressure Gradient, but is exactly equal to the static mud density required in the borehole to balance formation pore pressure. This term was first defined by Exploration Logging and is standard in all programs and logs.

- Normal Formation Balance Gradient is the normal hydrostatic pressure gradient measured from the flowline. The following examples illustrate the particular relationships between these gradients.

| DEPTH (FT) | | | | INTERVAL | FLUID DENSITY | | HYDROSTATIC PRESSURE | TOTAL HYDROSTATIC PRESSURE | FORMATION BALANCE GRADIENT | | |
|---|---|---|---|---|---|---|---|---|---|---|---|
| RKB | | Below Flowline | | | | | | | | | |
| From | To | From | To | (FT) | G/CC | LB/GAL | (PSI) | (PSI) | PSI/FT | LB/GAL | |
| (1) | (2) | (3) = (1) - $D_{FL}$ | (4) = (2) - $D_{FL}$ | (5) = (4) - (3) | (6) | (7) = 8.34 * (6) | (8) = 0.0519 * (5) * (7) | (9) = Σ (8) | (10) = (9) ÷ (4) | (11) = (10) ÷ 0.0519 | |
| RKB (0) | FL (9) | -- | -- | 9 | -- | -- | -- | -- | -- | -- | |
| FL (9) | Water Table (280) | 0 | 271 | 271 | -- | -- | -- | -- | -- | -- | |
| 280 | 700 | 271 | 691 | 420 | 1.00 | 8.34 | 182 | 182 | 0.263 | 5.1 | FRESH |
| 700 | 1000 | 691 | 991 | 300 | 1.00 | 8.34 | 130 | 312 | 0.314 | 6.1 | |
| 1000 | 1500 | 991 | 1491 | 500 | 1.01 | 8.42 | 219 | 530 | 0.356 | 6.9 | BRACKISH |
| 1500 | 2000 | 1491 | 1991 | 500 | 1.01 | 8.42 | 219 | 749 | 0.376 | 7.2 | |
| 2000 | 3000 | 1991 | 2991 | 1000 | 1.01 | 8.42 | 437 | 1186 | 0.397 | 7.6 | |
| 3000 | 3800 | 2991 | 3791 | 800 | 1.01 | 8.42 | 350 | 1536 | 0.405 | 7.8 | |
| 3800 | 4000 | 3791 | 3991 | 200 | 1.14 | 9.51 | 99 | 1634 | 0.410 | 7.9 | SALINE |
| 4000 | 4500 | 3991 | 4491 | 500 | 1.14 | 9.51 | 247 | 1881 | 0.419 | 8.1 | |

Figure 3-7. Normal FBG calculation worksheet

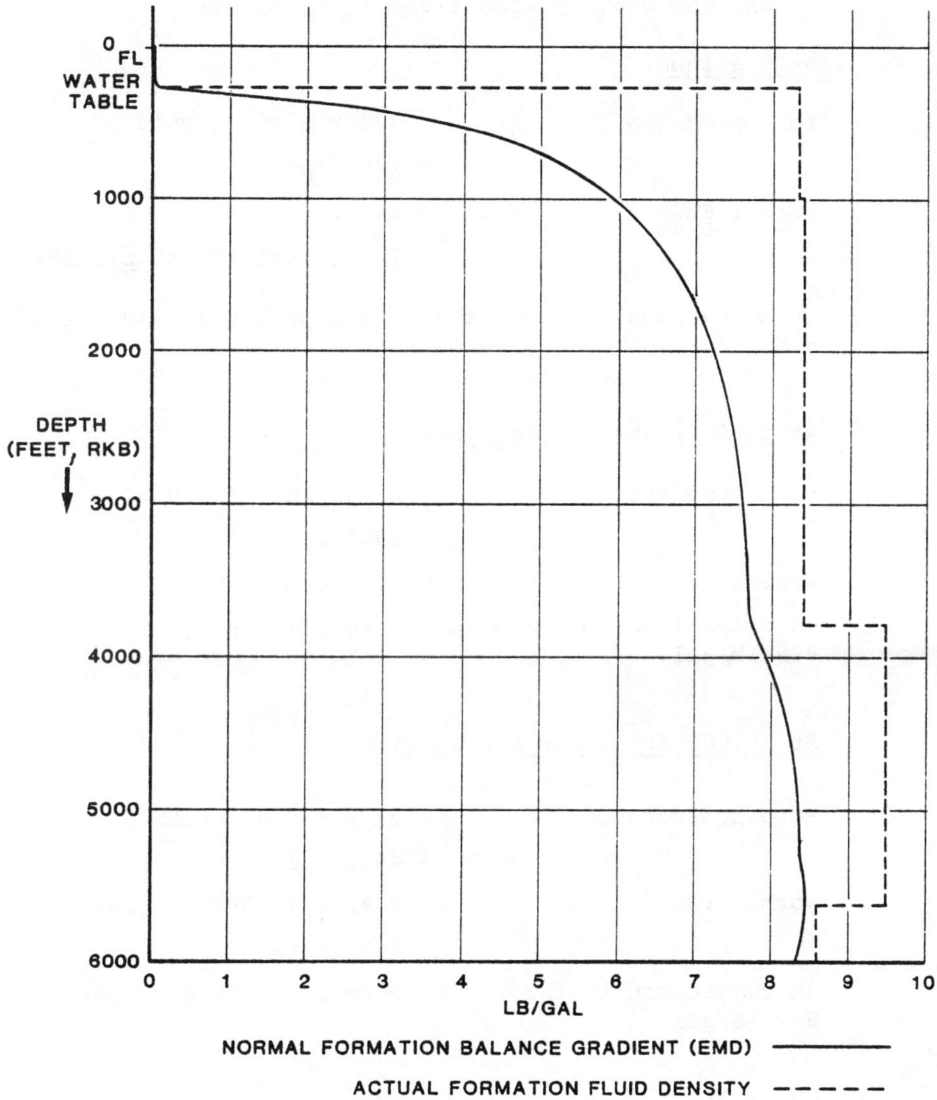


Figure 3-8. Actual formation fluid density and normal FBG (data from Figure 3-7)

Assume: Pore water density = 1.06 g/cc = 8.8 lb/gal;
offshore rig, water depth = 320 ft
air gap = 60 ft (sea level to flowline)

Then, <u>At seafloor</u>:

Pore pressure = 320 * 8.8 * 0.0519

= 146.2 psi

Normal FBG = 146.2 ÷ 380 ÷ 0.0519

= 7.4 lb/gal (0.385 psi/ft)

in comparison to the actual fluid pressure gradient of 8.8 lb/gal.

<u>At 3000 ft (below flowline)</u>:

Pore pressure = 2940 * 8.8 * 0.0519

= 1343 psi

Normal FBG = 8.6 lb/gal

in comparison to the actual pore pressure gradient of 8.8 lb/gal.

<u>At 10,000 ft (below flowline)</u>:

Pore pressure = 9940 * 8.8 * 0.0519

= 4540 psi

Normal FBG = 4540 ÷ 10,000 ÷ 0.0519

= 8.7 lb/gal

in comparison to the actual pore pressure gradient of 8.8 lb/gal.

With depth, it is thus apparent that the Normal Formation Balance Gradient will approach the actual pore pressure gradient asymptotically. In the above case, as the pore pressure gradient remains constant (equal to hydrostatic), the Normal Formation Balance Gradient is the same as the equivalent mud density that will precisely balance the pore pressure at any point. Section 4, paragraph 4.8, Drilling Exponents, shows the importance of this relationship.

Using the same rig situation, but with geopressures:

<u>At 3000 ft (below flowline)</u>:

| | |
|---|---|
| Pore pressure gradient | = 10.5 lb/gal (see Section 4) |
| Pore pressure | = 2940 * 10.5 * 0.0519 |
| | = 1602 psi |
| FBG | = 1602 ÷ 3000 ÷ 0.0514 |
| | = 10.3 lb/gal |

<u>At 10,000 ft (below flowline)</u>:

| | |
|---|---|
| Pore pressure gradient | = 10.5 lb/gal |
| Pore pressure | = 9940 * 10.5 * 0.0519 |
| | = 5417 psi |
| FBG | = 5417 ÷ 10,000 ÷ 0.0519 |
| | = 10.4 lb/gal |

Hence, in these cases, the Formation Balance Gradient equals the Equivalent Mud Density, but the Normal Formation Balance Gradient remains the same as in the first example, i.e. 7.4 lb/gal, 8.6 lb/gal and 8.7 lb/gal at seabed, 3000 feet and 10,000 feet, respectively. This is shown schematically in Figure 3-5.

The formation balance gradient at any point in the hole is actually measured as EMD. It is thus necessary to convert this gradient to a pressure (in psi or its metric equivalent) by the use of Equation (3-1) or (3-6).

Fracture pressures (see Section 5) may also be converted to equivalent mud densities. However, since fracture pressures vary considerably with changing lithology and pore pressures, the term "fracture pressure gradient" becomes almost meaningless. Nonetheless, at any point in the hole, the calculated fracture pressure can be converted to EMD; thus, this represents the mud density necessary to cause that pressure at that depth. By converting fracture pressure to EMD, convenience is gained — particularly for immediate well planning — but it should be remembered that equivalent mud density is a gradient referring to the mud in the hole, and not a property of the formation (see Figure 3-9).

## 3.6 EFFECTIVE OVERBURDEN PRESSURE

The effective overburden pressure is that portion of the overburden pressure that is not supported by the pore pressure. The effective overburden pressure is calculated by

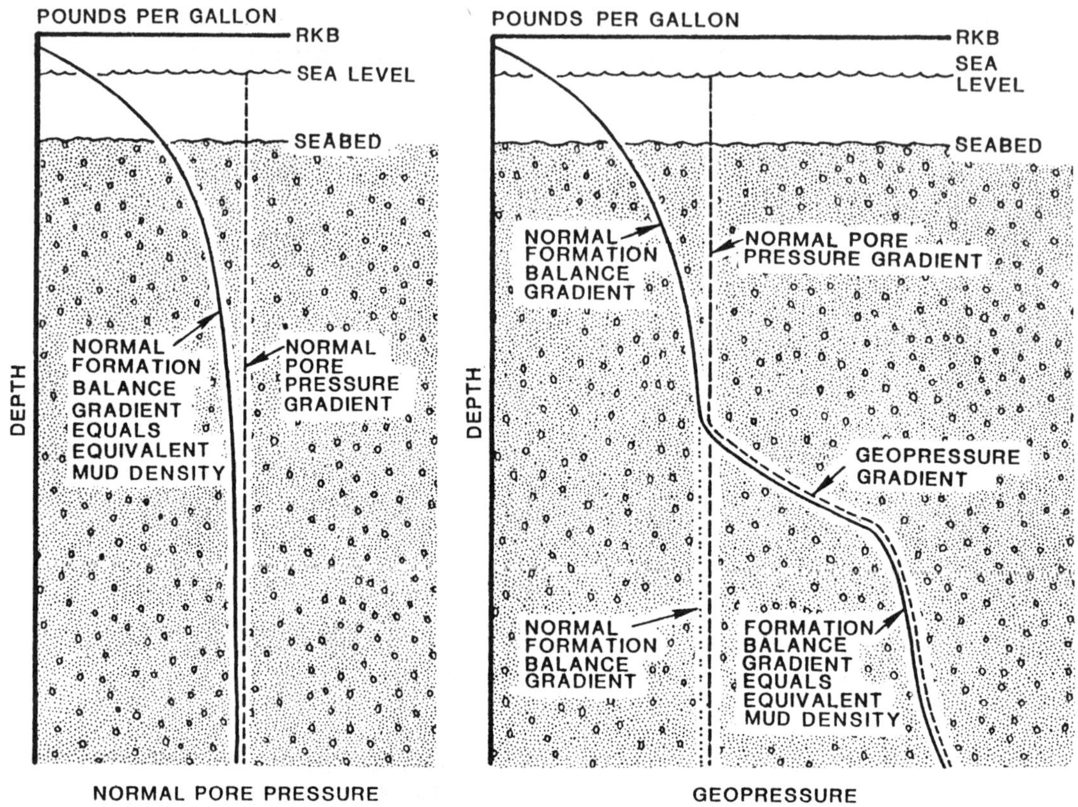


Figure 3-9. Actual formation fluid density and normal FBG

$$\sigma_1' = S - P \tag{3-12}$$

where

$\sigma_1'$ = effective overburden pressure (psi)

S = total overburden pressure (psi)

P = pore pressure (psi)

The term $\sigma_1'$ has no application in geopressure evaluation apart from fracture pressure calculation (Section 5, paragraph 5.16); nevertheless, it is important in understanding the relationship between pore pressures and overburden pressures.

As the pore pressure increases, more of the overburden becomes supported — reducing the effective overburden pressure. When the pore pressure is equal to the overburden pressure, the effective overburden pressure is zero; and when this

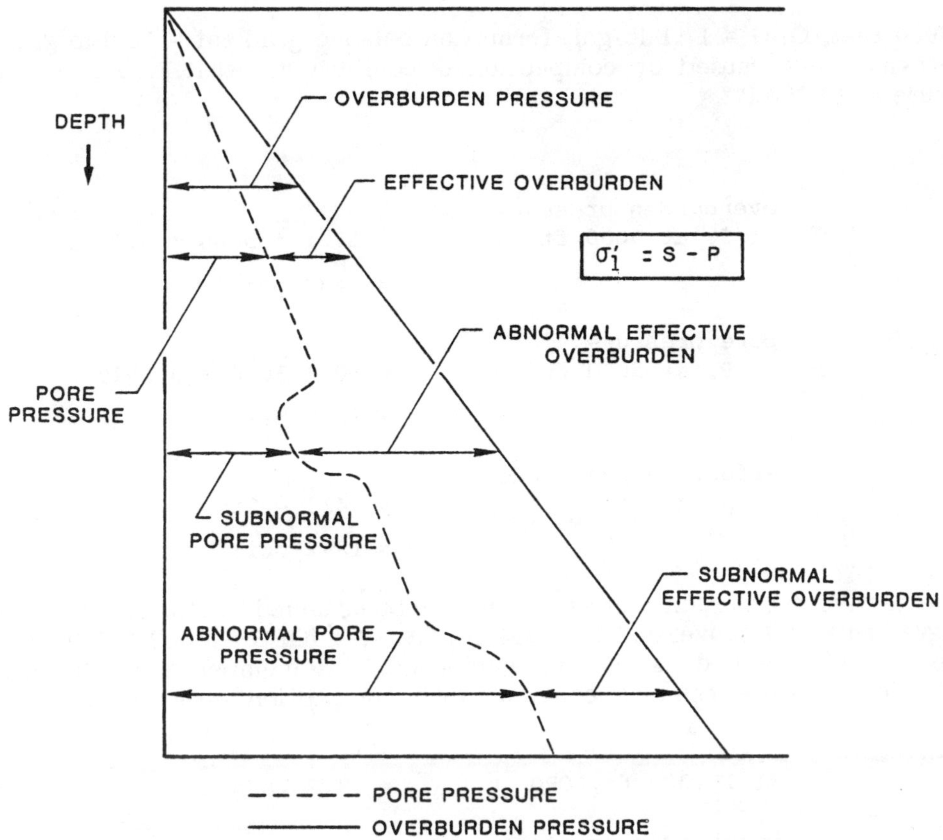


Figure 3-10. Effective overburden pressure in normal and geopressured formations

occurs, gravity sliding, diapirism, and other induced deformation may occur. The effective overburden pressure is that pressure which causes compaction. Therefore, compaction may still occur in geopressured zones, albeit at a slowed rate, unless the pore pressure is equal to the overburden pressure (see Figure 3-10).

If a geopressured zone is thought to be caused by compaction disequilibrium (Section 2, paragraph 2.2), pore pressures increase at the same rate as the overburden pressure, hence the effective overburden pressure remains constant. The expected rate of pore pressure increase can be calculated as follows:

$$\sigma_1' = S - P \qquad (3\text{-}12)$$

$$P = S - \sigma_1' \qquad (3\text{-}13)$$

where

$\sigma_1'$ = effective overburden pressure

S = overburden pressure

P = pore pressure

At 5000 feet, OBG = 17.1 lb/gal, formation balance gradient = 10.0 lb/gal. If the geopressure was caused by compaction disequilibrium, what would be the pore pressure at 10,000 ft?

| | |
|---|---|
| overburden pressure, S, at 5000 ft | = 17.1 * 5000 * 0.0519 |
| | = 4437 psi |
| pore pressure, P, at 5000 ft | = 10 * 5000 * 0.0519 |
| | = 2595 psi |
| effective overburden pressure $\sigma_1'$ | = 4437 - 2595 |
| | = 1842 psi |

(The effective overburden pressure could also be obtained by simply subtracting the two gradients and converting to pressure; i.e. $\sigma_1'$ = (OBG - FBG) * 5000 * 0.0519. However, this method is not recommended. Remember that the effective overburden pressure remains constant with compaction disequilibrium, not the gradient.)

| | |
|---|---|
| At 10,000 ft, OBG | = 18.2 lb/gal |
| overburden pressure, S, at 10,000 ft | = 18.2 * 10,000 * 0.0519 |
| | = 9446 psi |

As the effective overburden pressure remained constant,

| | |
|---|---|
| pore pressure, P at 10,000 ft | = 9446 - 1842 |
| | = 7604 psi |
| formation balance gradient | = 7604 ÷ 10000 ÷ .0519 |
| | = 14.7 lb/gal |

## 3.7 EFFECTIVE CIRCULATING DENSITY

In order to make full use of the formation pressures determined in pressure evaluation work, it is essential that the pressures existing in and imposed by the mud circulating system be known and fully understood.

The density of the drilling fluid itself does not remain constant throughout its cycle. For example, the weight of suspended cuttings in the annulus normally increases the effective density of the mud and hence the hydrostatic pressure imposed at the bottom of the hole.

An important factor in consideration of true bottomhole pressure is the effective backpressure imposed on the bottom due to the annular pressure loss. When circulating normally through an open flowline, measured mud pressure at the surface will be zero (casing pressure). However, frictional effects in the annulus present a restriction to flow, and a certain amount of pump pressure is required to overcome this restriction. This restriction acts in the same way as a closed-in choke applying a backpressure on the bottom in addition to hydrostatic pressure. The total pressure effective at the bottom during circulation is termed the bottomhole circulating pressure (BHCP), and its equivalent mud density is termed the Effective Circulating Density (ECD).

The extent of the flow restriction and hence the pressure loss is dependent upon total depth, annular dimensions, fluid viscosity, and flow regime, i.e., laminar or turbulent. Using the conventional Bingham model for a drilling fluid, the pressure losses may be approximated by

$$\delta p = \frac{L * YP}{A * (I.D. - O.D.)} + \frac{PV * L * V}{B * (I.D. - O.D.)^2} \qquad (3\text{-}14)$$

where

$\delta p$ = annular pressure loss (psi)

L = measured length of section (ft)

YP = yield point ($lb/100\ ft^2$)

I.D. -O.D. = hole (or casing) I.D. minus pipe (or collar) O.D. (inches)

PV = plastic viscosity (centipoise; cp)

V = annular velocity (ft/min)

A = 225 for drillpipe, 200 for annulus

B = 90,000 for drillpipe, 60,000 for annulus

This equation gives pressure losses in a pipe or annulus containing fluid moving in laminar flow, and tends to give slightly inflated values.

$$\underset{(ft/min)}{\text{annular velocity}} = \frac{24.51 * \text{gallons per minute}}{(I.D.^2 - O.D.^2)} \qquad (3\text{-}15)$$

When using tapered strings or in partially cased hole, the total pressure loss will be the sum of those calculated for the individual annular segments:

$$\text{ECD} = W + \frac{\Sigma\delta p}{0.0519 * D} \tag{3-16}$$

where

ECD = effective circulating density (lb/gal)

$\Sigma\delta p$ = total annular pressure loss (psi)

W = mud density (lb/gal)

D = vertical depth (ft)

$$\begin{aligned} \text{BHCP} &= \Sigma\delta P + (W * D * 0.0519) \\ &= \text{ECD} * D * 0.0519 \end{aligned} \tag{3-17}$$

where

BHCP = bottomhole circulating pressure (psi)

Notice that in calculating pressure losses the actual measured length of the Flow path is used. The sum of these will be the total measured depth of the well. When converting this pressure loss to an equivalent mud density (Equation 3-16), the vertical depth must be used since a hydrostatic column of fluid is being considered.

Using the Power Model for the behavior of a drilling fluid, annular pressure loss may be defined as

$$\delta_p = \frac{L\tau}{300\ (\text{I.D.} - \text{O.D.})} \tag{3-18}$$

where

L = measured length of section (ft)

$\delta_p$ = annular pressure loss (psi)

$\tau$ = shear stress (lb/100 $ft^2$)

I.D. −O.D. = hole (or casing) I.D. minus pipe (or collar) O.D. (inches)

Since the Power Model usually approximates more closely to true fluid behavior, this formulation may produce a more accurate annular pressure loss.

When pipe is pulled from open hole, the bottomhole pressure will be reduced due to the swabbing action caused by upward pipe movement. As the pipe moves upward, the friction between the pipe, mud and borehole wall produces a pressure reduction. The maximum effect on mud density is immediately below the bit; the maximum overall pressure reduction occurs at the bottom of the hole, due to the 'plunger' effect of the bit. An open drillstring allows fluid to flow through the bit, allowing some degree of pressure-relief, but if the drillstring has a float or downhole B.O.P., swab pressures will be at a maximum. As a very general rule of thumb, this pressure reduction may be of at least the same magnitude as the annular pressure loss; actual values depend on pipe pulling speeds and hole condition:

$$W_{trip} \leq W - \frac{\Sigma\delta p}{0.0519 * D} \qquad (3\text{-}19)$$

This pressure reduction due to swabbing can be vital when drilling geopressured intervals, as the lowering of the effective mud density may allow the well to flow.

See Figure 3-11 for a typical swab/surge printout from GEMDAS and Pressure Evaluation programs.

Large changes in mud density or effective mud density should be avoided. Sometimes a change may be brought about that is unexpected in magnitude and may lead to severe hole problems. One major instance is when surface casing is run on offshore wells, necessitating (in some cases) removal of the riser. The following example illustrates a possible series of events.

A floating rig is in 250 ft of water. The air gap is 45 ft, RKB to flowline is 5 ft, and 30-inch casing was set at 600 ft. The B.O.P. and riser were installed, and hole was drilled to the 20-inch casing point at 1500 ft. High gas shows occurred from 800 and 1100 ft with the mudweight at 9.5 lb/gal. (Refer to Section 4 for importance of shallow gas zones.) In order to run 20-inch casing, it is necessary to pull the riser. With 9.5 lb/gal mud in the hole, the following pressures are present:

```
hydrostatic pressure, at 600 ft = 9.5 *  (600-5) * 0.0519 = 293 psi
                  ... at  800 ft = 9.5 *  (800-5) * 0.0519 = 392 psi
                  ... at 1100 ft = 9.5 * (1100-5) * 0.0519 = 540 psi
                  ... at 1500 ft = 9.5 * (1500-5) * 0.0519 = 737 psi
```

In order to pull the riser, it is first necessary to displace it with seawater (density 8.5 lb/gal). When this is done, the resultant pressures would be:

```
At seabed, hydrostatic pressure = (250 + 45 - 5) * 8.5 * 0.0519 = 128 psi
                  ... at  600 ft =  (600-295)*9.5*0.0519  + 128 = 278 psi
                  ... at  800 ft =  (800-295)*9.5*0.0519  + 128 = 377 psi
                  ... at 1100 ft = (1100-295)*9.5*0.0519  + 128 = 525 psi
                  ... at 1500 ft = (1500-295)*9.5*0.0519  + 128 = 722 psi

Resulting gradients of EMD are 9.0 lb/gal at  600 ft
                               9.1 lb/gal at  800 ft
                               9.2 lb/gal at 1100 ft
                               9.3 lb/gal at 1500 ft
```

IRWELL VALLEY DRILLING CONSORTIUM : TEST WELL # 1
3:40 5/ 9/81

SWAB AND SURGE ANALYSIS

TOTAL DEPTH (ft) = 15000.0, CASING SHOE DEPTH =8000.0, CLOSED PIPE
MUD WEIGHT = 12.00 lb/gal, PLASTIC VISCOSITY = 28.00, YIELD POINT = 14.00
LOW RANGE POWER LAW: K = .452, N = .737
MID RANGE POWER LAW: K = .452, N = .737

| RUNNING SPEED | | BIT ON BOTTOM | | | BIT AT SHOE | | |
|---|---|---|---|---|---|---|---|
| ft/mi | SEC/STND | psi | TD EQUIV MUD WT SURGE | TD EQUIV MUD WT SWAB | psi | SHOE EQUIV MUD WT SURGE | SHOE EQUIV MUD WT SWAB |
| 558.00 | 10 | 675.7 | 12.91 | 11.09 | 403.2 | 12.98 | 11.02 |
| 279.00 | 20 | 213.8 | 12.29 | 11.71 | 127.6 | 12.31 | 11.69 |
| 186.00 | 30 | 89.0 | 12.12 | 11.88 | 53.6 | 12.13 | 11.87 |
| 139.50 | 40 | 65.5 | 12.09 | 11.91 | 37.7 | 12.09 | 11.91 |
| 111.60 | 50 | 55.5 | 12.07 | 11.93 | 32.0 | 12.08 | 11.92 |
| 93.00 | 60 | 48.6 | 12.07 | 11.93 | 28.0 | 12.07 | 11.93 |
| 79.71 | 70 | 43.3 | 12.06 | 11.94 | 25.0 | 12.06 | 11.94 |
| 69.75 | 80 | 39.3 | 12.05 | 11.95 | 22.6 | 12.05 | 11.95 |
| 62.00 | 90 | 36.0 | 12.05 | 11.95 | 20.7 | 12.05 | 11.95 |
| 55.80 | 100 | 33.3 | 12.04 | 11.96 | 19.2 | 12.05 | 11.95 |
| 50.73 | 110 | 31.1 | 12.04 | 11.96 | 17.9 | 12.04 | 11.96 |
| 46.50 | 120 | 29.1 | 12.04 | 11.96 | 16.8 | 12.04 | 11.96 |
| 42.92 | 130 | 27.5 | 12.04 | 11.96 | 15.8 | 12.04 | 11.96 |
| 39.86 | 140 | 26.0 | 12.04 | 11.96 | 15.0 | 12.04 | 11.96 |
| 37.20 | 150 | 24.7 | 12.03 | 11.97 | 14.2 | 12.03 | 11.97 |
| 34.88 | 160 | 23.6 | 12.03 | 11.97 | 13.6 | 12.03 | 11.97 |
| 32.82 | 170 | 22.5 | 12.03 | 11.97 | 13.0 | 12.03 | 11.97 |
| 31.00 | 180 | 21.6 | 12.03 | 11.97 | 12.4 | 12.03 | 11.97 |
| 29.37 | 190 | 20.8 | 12.03 | 11.97 | 12.0 | 12.03 | 11.97 |
| 27.90 | 200 | 20.0 | 12.03 | 11.97 | 11.5 | 12.03 | 11.97 |

Figure 3-11. ELOS Swab and Surge Calculation

Note that the gradients at 800 and 1100 ft (9.1 and 9.2 lb/gal, respectively) are now much less than the original 9.5 lb/gal used when drilling. If these zones are permeable gas zones of between 9.0 and 9.5 lb/gal formation balance gradient, a problem may result when the riser is disconnected. When the riser is disconnected, the fluid level in the riser falls to sea level, causing further pressure reduction:

At seabed, the hydrostatic pressure = 250 * 8.5 * 0.0519 = 110 psi

pressure at 600 ft = (600-295)*9.5*0.0519 + 110 = 260 psi

... 800 ft = (800-295)*9.5*0.0519 + 110 = 359 psi

...1100 ft = (1100-295)*9.5*0.0519 + 110 = 507 psi

...1500 ft = (1500-295)*9.5*0.0519 + 110 = 704 psi

Resulting gradients of EMD are 8.4 lb/gal at 600 ft

8.7 lb/gal at 800 ft

8.9 lb/gal at 1100 ft

9.1 lb/gal at 1500 ft

Note that the reduction of only 18 psi throughout the column, caused by disconnecting the riser, lowered the gradients sufficiently to create major underbalance. The zones at 800 and 1100 ft may flow, and if the riser is disconnected, control of the well would be extremely difficult.

In order to keep a 9.5 lb/gal gradient at 1100 ft, it is necessary to weight up the mud in the hole before disconnecting the riser. The new mud density may be calculated as follows.

$$\text{New mud density} = \frac{(D * W) - 8.5\,(Dw - BOP_L)}{D - Dw - A + BOP_L} \qquad (3\text{-}20)$$

where

D = vertical depth of hole (ft, from flowline)

W = mud density in the hole (lb/gal)

Dw = water depth (ft)

$BOP_L$ = height of BOP stack from seabed to riser connector (ft)

A = distance from flowline to sea level (ft)

8.5 = density of seawater (lb/gal)

Using the above example where the height of the BOP stack is 35 ft, in order to keep 9.5 lb/gal gradient at 1100 ft, the new mud density must be

$$W = \frac{(1095 * 9.5) - 8.5\,(250 - 35)}{1095 - 250 - 40 + 35}$$

$$= 10.2 \text{ lb/gal}$$

This increase in mud density, or riser margin, must be known at all times as the well is being drilled. Should a situation arise whereby it becomes necessary to move off location (e.g., storms, ice movements, rig damage, etc.), the geologist should be able to provide the operator with the riser margin whenever required to do so. It is important to note that the riser margin in shallow sediments and very deep water may be too high; the mud density increase cannot be circulated, as the minimum formation fracture pressure may be exceeded. In these situations a rig may have two risers, one for drilling top hole, and when surface casing has been set, the riser is exchanged for the narrower one. It may be necessary, however, to attempt to drill surface hole without a riser, but this can be hazardous if shallow gas is encountered.

## 3.8 DIRECT PRESSURE MEASUREMENTS

Downhole pressure bombs, DSTs, RFTs and FITs can all give accurate reservoir pressures which are reliable indicators for the planning of future wells, but are rarely used to determine the magnitude of geopressures because of the

- difficulty in the location of permeable formations in overpressured zones
- risk of differential sticking
- risk of collapse of impermeable formations
- high cost of rig time and tools

A method of observing true formation pressure without special equipment is by recording surface pressures when a well has kicked and is shut in. While "drilling for a kick" is no longer widely accepted as a viable drilling procedure, kicks are occasionally taken — and it is necessary to calculate formation pressure in order to kill the well.

When a well is shut in there is a static system of three pressures in balance. Since the well cannot flow, formation pressure is known to be balanced. Since the pressures recorded at the standpipe (shut-in drillpipe pressure, SIDP) and at the casing head (shut-in casing pressure, SICP) have stabilized and no external pressure is being applied at the surface, it can be deduced that the formation pressure is exactly in balance. Thus,

$$P = P_{md} + SIDP = P_{ma} + P_k + SICP \tag{3-21}$$

where

$P$ = formation pore pressure (psi)

$P_{ma}$ = hydrostatic pressure of the mud in the annulus (psi)

$P_{md}$ = hydrostatic pressure of the mud in the drillstring (psi)

$P_k$ = hydrostatic pressure of kick fluid present in the annulus (psi)

SIDP = shut-in drillpipe pressure (psi)

SICP = shut-in casing pressure (psi)

Since the mud density, total depth, and shut-in drillpipe pressure are known, it is possible to substitute Equation (3-1) in Equation (3-21) and solve for formation pore pressure:

$$P = (0.0519 * D * W) + SIDP \tag{3-22}$$

Hence,

$$W' = W + \frac{SIDP}{0.0519 * D} \tag{3-23}$$

where

D = vertical depth (ft)

W = mud density (lb/gal)

W' = mud density to kill the well, i.e., formation balance gradient (lb/gal)

See Section 5, paragraph 5.23, for a more complete analysis.

## 3.9 LOG-DERIVED FLUID DENSITIES

The spontaneous potential (SP) log is a recording of the difference between the potential of a moveable electrode in the borehole and the fixed potential of a surface electrode. It was the first log type used in the petroleum industry. The SP log is used to

- Identify permeable formations
- Correlate geological horizons
- Determine qualitative shaliness
- Determine formation water resistivity, Rw

The SP curve in shales shows generally consistent values and follows a straight line; this is the shale baseline and is the base from which deflections in permeable formations are read. Opposite permeable formations, the SP shows deflections from the shale baseline, and in thick beds the trace tends to remain at an essentially constant deflection: this is the static SP (SSP). In thinner beds the SP may not reach its full deflection, hence a bed-thickness correction is applied to estimate the SSP. The deflections from the shale baseline may be either to the left (negative) or to the right (positive), depending on the relative salinities of the formation water and the mud filtrate.

The position of the shale baseline has no meaning for interpretation purposes. The baseline position is set by the engineer so that all the deflections remains inside the SP track.

There are eight major points that influence the shape and amplitude of an SP deflection:

- Formation thickness (h)
- Formation true resistivity ($R_t$)
- Resistivity of the mud ($R_m$) and diameter of the borehole ($d_h$).
- Resistivity ($R_{xo}$) of the invaded zone, and diameter of the invaded zone ($d_i$)
- Resistivity of the adjacent formations ($R_s$)
- Shaliness of the formation
- Shale baseline shifts (polarization, manual)
- Noise

The experienced log analyst can take all these considerations into account quantitatively; however, for our purpose, an awareness of the various influencing parameters is sufficient in order to obtain resultant values of adequate accuracy.

By calculating $R_w$ for various water-bearing permeable zones, it is possible to accurately define the density of these pore waters by the use of simple conversion charts. This data is thus invaluable for providing precise normal hydrostatic gradients. (Refer to Appendix B for the relevant charts.)

In brief, the procedure for determining fluid density from electric log data is as follows:

1) Determine the diameter of invasion from the appropriate $R_{int}$ chart (Schlumberger, "Log Interpretation Charts"). It will frequently be a sufficient approximation to assume no invasion, however.

2) Correct the SP reading for bed thickness and invasion, using chart B-1 or B-2. The results may be considered reliable only if the correction does not exceed 20%.

3) Convert Rmf at 75° F to Rmf at formation temperature, using nomogram B-3.

4) Convert Rmf to $Rmf_{eq}$. If Rmf at 75° F exceeds 0.1 ohm-m, use $Rmf_{eq}$=0.85 Rmf (at formation temperature). Otherwise, use chart B-4 for the conversion.

5) Find $Rw_{eq}$, using chart B-5.

6) Convert $Rw_{eq}$ to Rw, using chart B-4.

7) Convert Rw to salinity, using nonomgram B-3.

8) Deterine fluid density, using chart C-1.

9) Calculate normal FBG-Depth relationship using correct rig elevations (see Figure 3-7).

In ELOS, an EAP action is available to give a rapid approximate solution, with sufficient accuracy for application in pressure evaluation (see Figure 3-12).

```
                                        IRWELL VALLEY DRILLING CONSORTIUM : TEST WELL # 1
                                         3:37        5/ 9/81
Rw DETERMINATION FROM THE SPONTANEOUS POTENTIAL

BOTTOM HOLE TEMPERATURE AT 10000.0 ft = 100.0 deg C
FORMATION TEMPERATURE AT SURFACE = 20.0 deg C
RESISTIVITY OF MUD FILTRATE, Rmf = .093 ohm sqm/m AT 25.0 deg C


AT DEPTH = 8000.0 ft
SPONTANEOUS POTENTIAL = -36.0 mv
TEMPERATURE (deg C) = 84.0
     RMFE = .003
     RMF  = .033
     RWE  = .001
     RW   = .031


AT DEPTH = 8200.0 ft
SPONTANEOUS POTENTIAL = -36.5 mv
TEMPERATURE (deg C) = 85.6
     RMFE = .003
     RMF  = .032
     RWE  = .001
     RW   = .031


AT DEPTH = 8400.0 ft
SPONTANEOUS POTENTIAL = -34.2 mv
TEMPERATURE (deg C) = 87.2
     RMFE = .003
     RMF  = .032
     RWE  = .001
     RW   = .030


AT DEPTH = 8600.0 ft
SPONTANEOUS POTENTIAL = -37.5 mv
TEMPERATURE (deg C) = 88.8
     RMFE = .003
     RMF  = .031
     RWE  = .001
     RW   = .030


AT DEPTH = 8800.0 ft
SPONTANEOUS POTENTIAL = -36.5 mv
TEMPERATURE (deg C) = 90.4
     RMFE = .003
     RMF  = .031
     RWE  = .001
     RW   = .029


AT DEPTH = 9000.0 ft
SPONTANEOUS POTENTIAL = -36.7 mv
TEMPERATURE (deg C) = 92.0
     RMFE = .003
     RMF  = .030
     RWE  = .001
     RW   = .029
```

Figure 3-12. ELOS SP log analysis

# 4
# PORE PRESSURE EVALUATION TECHNIQUES

## 4.1 INTRODUCTION

Drilling into a geopressured zone causes a change in a number of basic formation/drilling relationships. This change is usually a reversal of a gradual depth-related trend in a lithologically uniform formation.

- Compaction increases uniformly with depth in a normal pressured clay rock. A geopressured zone may be poorly compacted relative to those zones overlying it.
- Porosity and water content decrease uniformly with depth in a normal pressured clay rock. A geopressured zone in which dewatering has been slowed will show a reversal in the trend, with increased water content and increased porosity.
- Other factors relating to fluid movement, such as ionic concentrations, hydrocarbon saturations, etc., may be different in geopressured zones.
- Differential pressure across bottom, which increases with depth when a normal pressured formation is drilled with a constant mud density, thus decreases or even reverses when a geopressured zone is penetrated.

Thus, any measurable parameter which reflects any or all of these factors may be used as a means of interpreting changes in formation pressure and eventually as a means of evaluating and obtaining quantitative estimates of formation pore pressures.

Remember, however, that these properties and the parameters that reflect them vary between lithologies, and that a drilling break or reversal of a trend may simply indicate that a lithological change has occurred and a new trend must be established. Similarly, minor lithological variations introduce minor variations in the individual parameters. Care should be taken in interpretation to account for these lithological variations.

Before the introduction of the Pressure Evaluation Log Suite and the specialized recording systems, there were a number of ways of detecting geopressures, as outlined in paragraphs 4.4, 4.6, 4.7 and 4.11. It is crucial not to rely on new methods to such an extent that good logging practice and experience are ignored. Similarly, when only a formation evaluation log is being plotted, the logging geologist should be constantly alert to the occurrence of these geopressure-indicative phenomena and, should they occur, report them immediately to the operator and make a note of suspect geopressure on the log.

Recognizing the existence of a geopressure is an essential first stage in overall well control. Alone, it serves as an excellent tool in well evaluation, economics and safety. However, for complete well control, it is necessary that not only the presence but also the magnitude of a pressure abnormality be known. Complete well control is an ideal that even with the best equipment and personnel cannot be reached. Drilling activity, lithological changes, and the type and history of a geopressure all affect the degree of accuracy with which its magnitude may be estimated. When reporting, the GEMDAS Operator or P.E.G. should always specify his confidence level and never be afraid to express uncertainty.

Pressure determinations by direct measurement have disadvantages previously enumerated and may be made only after a pressure abnormality has been entered and a permeable zone encountered. These methods are therefore severely limited for real-time well planning, although they may be of value in preparing future well prognoses.

In areas where sufficient data is available, it is possible to prepare correlation charts relating trend deviation to known formation pressure data. These charts may then be used for future wells, to estimate pressures from trend deviations. However, they are reliable only in the area for which they were prepared, which may be extremely localized. Minor variations within the area result in the pressure determination being a vague, qualitative estimate at best, and attempts to use a chart outside its area of preparation, even in an area of similar geological setting, can be literally disastrous.

Attempts to use cap-rock deviation as a pressure indicator have rarely proved to be of value. In addition to the limitations on the geopressure deviation methods, the efficiency of the "seal" may be affected by mineralogical variations and by vertical extent; e.g., a thick seal of moderate permeability may be as efficient as a thin seal of low permeability. Furthermore, the magnitude of the pressure abnormality contained by the seal will be dependent upon the overall thickness of sediments within the sealed zone, the presence of flow conduits below the seal, and the age of the formations. In the cases where correlation has been possible between cap-rock deviation and known pressure data, there was a thin, discrete cap rock above a relatively uniform pressure abnormality. Such cases have proved to be severely restricted geographically. While the collation of such data should be carried out, and may prove of value in certain areas, little faith should be placed in the method.

Before a new well commences, a prespud meeting may be arranged with the client. During this meeting all relevant data from nearby wells, seismic anomalies, and geological data should be collected and discussed in order that suspected problem zones can be delineated and analyzed. This opportunity should also be used to ascertain communication channels and reporting procedures, and to review the drilling prognosis to ensure that suitable measures are planned in the event of encountering geopressures.

## 4.2 GEOPRESSURE EVALUATION FROM SEISMIC DATA

The success and accuracy to which geophysicists can predict formation boundaries in the subsurface through seismic interpretation has been used to great advantage in determining possible hydrocarbon provinces. Formations to 20,000 ft depth can be delineated to within 98 percent accuracy, but with increasing depth this accuracy deteriorates rapidly. However, with a different geophone spread, greater resolution (better than 1 percent error) can be consistently obtained for predicting formation tops below 20,000 ft. The highest accuracy can be consistently maintained in an area in which the geology is relatively well known; for example, in a Tertiary section of simple sand/shale sequence, seismic data can not only predict formation boundaries, but subtle reflections allow interpretation of small fault movements, unconformities and geopressured intervals. In rank wildcat areas, the lack of data on subsurface lithologies and geologic age is a major handicap in interpreting even formation boundaries. However, construction of many velocity curves with depth from the surrounding area should adequately delineate the normal compaction trend for the area, i.e., normal velocity increase with depth.

The seismic data that is used mostly for geopressure indications is velocity analysis. The resultant curves are representative of the sonic velocity of the formations to be drilled in a gross sense. Normal changes in lithology are not delineated by these curves; however, in the majority of cases, geopressured intervals may be recognized. Again, knowledge of the geology markedly increases accuracy, as a limestone/dolomite sequence overlying a thick clay interval will show the characteristic velocity reversal which also occurs across the normally-pressured/geopressured transition in shale. Figure 4-1 shows typical velocity analyses for different lithological sections. A very similar curve is produced by the

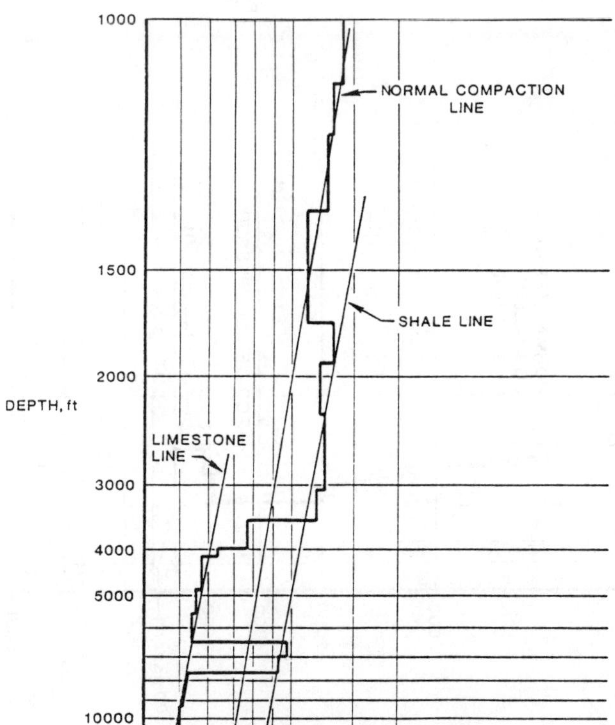


Figure 4-1. Interval transit time variations with compaction and lithology

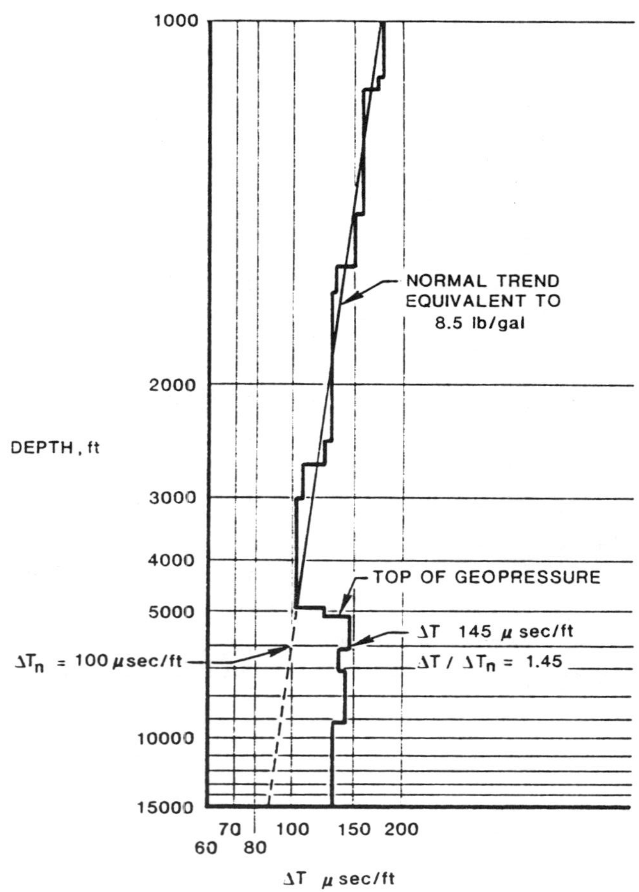


Figure 4-2. Interval transit time variation with formation pore pressure

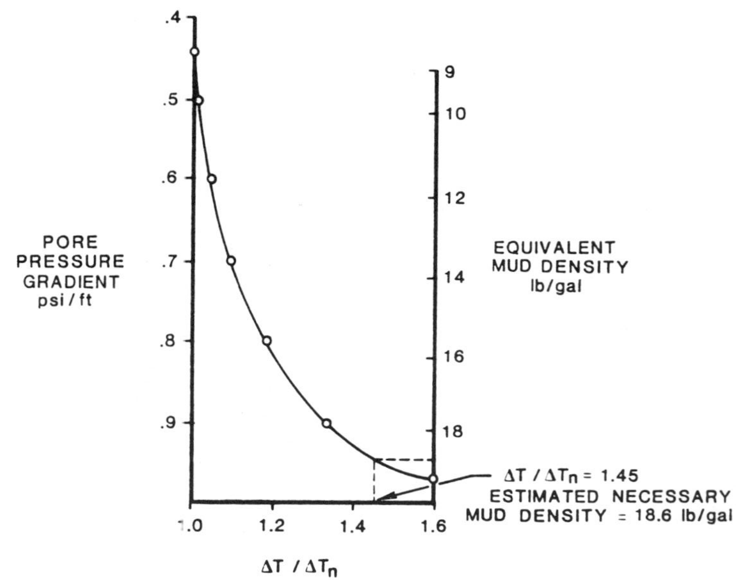


Figure 4-3. Geopressure evaluation from interval transit time

velocity survey in a borehole. Velocity is usually translated into transit time for the convenience of subsequent correlation with the sonic log, but the interpretation remains essentially the same except that an increase in transit time is synonymous with a decrease in velocity.

Since sound velocity through a material is mainly dependent upon its elasticity and density, but can also be considerably modified by porosity, pore geometry, and other anisotropies, the normal response is for velocity to increase with depth. Departure from this normal trend is thus due either to gross lithological changes or geopressure, which specifically results in a departure to lower velocities with depth. In a rank wildcat well it is best to assume that any departure to lower velocities with depth is due to geopressure, so that the well may be safely planned accordingly. In areas of well known geology, a geopressured zone can be recognized with a far greater degree of certainty, as the lithological characteristics would be known. Pennebaker (1968) indicates that if the rock type remains constant, i.e., uniform clays, the degree of a departure to lower velocities is directly related to the increase in pore pressure. Figure 4-2 shows this relationship. A calibration curve, shown in Figure 4-3, was developed for Gulf Coast wells, but for broad estimates it should suffice for other similar Tertiary basins.

To estimate the formation balance gradient from velocity analysis, extrapolate the normal trend developed in hydrostatically pressured formations. At the depth of interest, determine the ratio of $T/T_n$ (if the velocity analysis is calibrated in transit time), or convert velocity to interval transit time by:

$$T = \frac{1}{V} * 1000 \qquad (4\text{-}1)$$

where

$T$ = interval transit time ($\mu$sec/ft)
$V$ = velocity (1000 ft/sec)

Use the resulting ratio with Figure 4-3 to obtain an estimate of the formation balance gradient at the depth of interest. Note that the velocity analyses (in $\mu$sec/ft) are plotted on a log-log grid: the normal compaction trend approximates a straight line on a log depth-scale, facilitating normal trend extrapolation.

Common sources of error in velocity analyses are due to dipping beds, faults, multiple reflections, curved ray paths, processing, and interpretation. Usually, the best quality velocity analyses are obtained from good quality seismic sections: good reflections give good root mean squared velocities (Reynolds, 1970).

## 4.3 DRILLING PARAMETERS

## 4.4 MUD DENSITY/GAS RELATIONSHIP

Differential pressure results from the difference between the ECD and the formation balance gradient. As shown in paragraph 3.7, it is normal and desirable to maintain the mud density slightly higher than the formation balance gradient. The resulting differential pressure may be calculated as follows:

$$(W * D * 0.0519) - (FBG * D * 0.0519) = \Delta P \qquad (4\text{-}2)$$

where

W = mud density (lb/gal)

D = depth (ft)

FBG = formation balance gradient (lb/gal)

$\Delta$P = differential pressure (psi)

Substituting ECD for W gives the differential pressure while drilling. $\Delta$P should be positive during all drilling operations, hence accurate pore pressure estimations are necessary.

Differential pressure is one of the major forces that affects the amount of gas that enters the mud, and is related to the amount of gas that is measured at the surface. By interpretation of the gas magnitude/formation/mud density relationships, a very good estimate of the formation balance gradient may be obtained. For example:

12-1/4-inch hole is being drilled at 2000 feet with a mud density of 9 lb/gal, and the formation balance gradient is 8.6 lb/gal.

$$\begin{aligned} P &= (9 * 2000 * 0.0519) - (8.6 * 2000 * 0.0519) \\ &= (9 - 8.6) * 2000 * 0.0519 \\ &= 42 \text{ psi} \end{aligned}$$

The same parameters at 15,000 ft:

$$\begin{aligned} P &= (9 - 8.6) * 15{,}000 * 0.0519 \\ &= 311 \text{ psi} \end{aligned}$$

Note that pressure differences in shallow hole are relatively small but nevertheless extremely important.

Clearly, the volume of gas released from a drilled formation is dependent upon the porosity, permeability, gas saturation, and differential pressure. Thus if the differential pressure is high, less gas will be released from a sand bed than from a clay bed if all variables (except the permeability) are the same. Conversely, if the differential pressure is low or even negative, far more gas will be released from a sand than from a clay with the same porosity, gas saturation and pore pressure, because the permeability is higher.

Negative differential pressure (while drilling) complicates interpretations as gas influx may be continually occurring. This is shown by increasing background gas, particularly when just circulating. Negative differential pressure while tripping may result in swabbing, a kick, or severely gas-cut mud upon recirculation. A very small or close-to-zero differential pressure may cause connection gases to be produced from permeable formations; however, connection gases produced from clays are indicative of reasonably high negative differential pressure.

Figure 4-4 demonstrates the effect of varying differential pressure on gas show magnitude. The total gas curves for two wells drilled through a similar section are shown. The data for both wells has also been normalized to reduce the effects of hole diameter, rate of penetration, mud pump output, and surface extraction efficiency. (This procedure is explained in Mud Logging: Principles and Interpretations (EXLOG, 1985). Well A was drilled using a constant mud density, whereas in well B mud density was controlled to maintain a constant positive differential pressure (overbalance).

In the upper portion of the section, the two gas curves are similar and the normalized gas curves overlay almost exactly. In the lower portion a progressive deviation between the two wells is seen which is somewhat reduced but remains evident even in the normalized curves. We can interpret this as being due to the penetration of a transition zone into a geopressure.

In Well A, maintaining a constant mud density results in a decreasing overbalance and eventually an underbalance or increasing negative differential pressure. Connection gases occur and become larger with deeper penetration. Additionally, feed-in of gas from the underbalanced borehole wall causes an increase in background gas which, since it is not a product of fresh-cut formation, cannot be accounted for in the normalization calculation.

Well B, on which a constant overbalance was maintained by increases in mud density, did not show increases in gas background or connection gases. Indeed, if any zone showed good permeability, the overbalance may have resulted in flushing gas away from the borehole and a reduction in observed total gas.

By careful observation of these various phenomena, a fairly accurate log of differential pressure (and hence pore pressure) may be obtained. This information should be used in conjunction with the other techniques described in the following paragraphs.

A large gas show in surface hole is indicative of very high porosity and gas saturation. Shallow gas does not expand very much before it reaches the surface, in comparison to gas from deep formations which expands enormously as it approaches the surface.

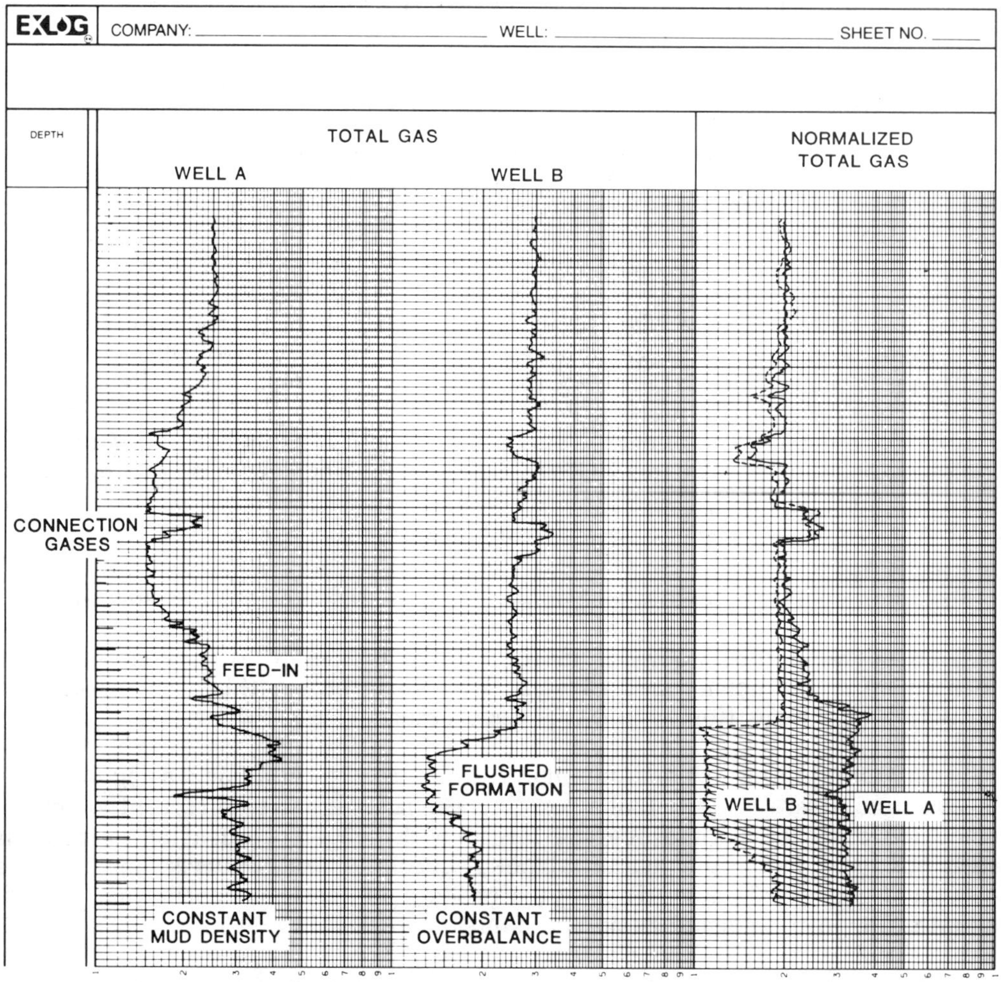


Figure 4-4. The effect of differential pressure on gas show magnitude

## 4.5 GAS-CUT MUD

Mud density reduction due to gas cutting is commonly not considered to be a cause for concern. It can, however, cause serious problems in top-hole.

Most of the gas that causes gas cutting is that liberated from the cuttings. As the cuttings are circulated up the hole, hydrostatic pressure becomes reduced, and the gas in the pores expands and is released to the mud. For example:

Gas entering the mud system (Goldsmith, 1972) is:

$$G_v = \left(\frac{d}{24}\right)^2 * \frac{\pi * R}{60} * \emptyset * Sg * 7.48 \qquad (4\text{-}3)$$

where

$G_v$ = rate of gas entering the mudstream at reservoir pressure (gal/min)

R = rate of penetration (ft/hr)

d = hole diameter (inches)

Ø = porosity (fractional)

$S_g$ = gas saturation (fractional)

For example:

$$V = \left(\frac{8.5}{24}\right)^2 * \frac{\pi * 85}{60} * 0.25 * 0.7 * 7.48$$

using d = 8.5

R = 85

Ø = 0.25

$S_g$ = 0.70

Reservoir pressure = 7000 psi

V = 0.731 gal/min at 7000 psi

The gas volume each minute at atmospheric pressure of 14.7 psi, using the ideal gas law and neglecting temperature effects, is:

$$G_{va} = G_v * \frac{P}{14.7} \qquad (4\text{-}4)$$

$$= 0.731 * \frac{7000}{14.7} = 348 \text{ gal/min at atm press}$$

Hence when the gas reaches the surface, the volume of gas flowing with the mud is about 350 gallons each minute. This gas, mixed with 280 gallons of mud each minute, results in a mud density of

$$W_1 = \frac{\text{mud (gpm)}}{\text{mud (gpm)} + \text{gas (gpm)}} * W_2 \tag{4-5}$$

$$= \frac{280}{280 + 350} * 9.2 \text{ lb/gal} = 4.1 \text{ lb/gal}$$

where

$W_1$ = gas-cut mud density (lb/gal)

$W_2$ = uncut mud density (lb/gal)

Increasing the mud density will not reduce this gas cutting, as the hydrostatic pressure of 9.2 lb/gal mud at 15,000 feet is 7162 psi, 162 psi greater than the pore pressure.

The decrease in bottomhole pressure caused by the most drastic gas cutting is negligible in deep wells, but can be a major problem in surface hole. For this reason large gas shows and concomitant mud cutting at shallow depth should be treated with the utmost caution. The pressure reduction caused by mud-cutting is given by (Goldsmith, 1972):

$$\Delta P = 14.7 \left( \frac{W_2 - W_1}{W_1} \right) \ln \left( \frac{3.53 * W_2 * D}{1000} \right) \tag{4-6}$$

where

$\Delta P$ = pressure reduction caused by mud cutting (psi)

$W_1$ = gas-cut mud density at the flowline (lb/gal)

$W_2$ = uncut mud density (lb/gal)

$D$ = depth of gas zone (ft)

$$\Delta P = 14.7 \left( \frac{9.2 - 4.1}{4.1} \right) \ln \left( \frac{3.53 * 9.2 * 15{,}000}{1000} \right)$$ (from the previous example

$$\Delta P = 113 \text{ psi}$$

Hence the actual mud gradient at 15,000 feet is

$$W = (7162 - 113) \div 15{,}000 \div 0.0519$$

$$= 9.1 \text{ lb/gal}$$

For gas-cut mud in shallow hole, however, the problem becomes greatly magnified:

Hole size is 12-1/4 inches, rate of penetration is 500 ft/hour, depth is 1000 feet. Formation has 30% porosity and 70% gas saturation, formation pore pressure is 467 psi (9 lb/gal), mud density is 9.2 lb/gal, and pump rate is 450 gal/min. Gas entering the mud system is:

$$\left(\frac{12.25}{24}\right)^2 * \left(\frac{\pi * 500}{60}\right) * 0.3 * 0.7 * 7.48 = 10.7 \text{ gal/min at 467 psi}$$

Gas volume each minute at atmospheric pressure is:

$$10.7 * \frac{467}{14.7} = 340 \text{ gal/min at atmospheric pressure.} \qquad (4\text{-}4)$$

Resultant mud density:

$$\frac{450}{450 + 340} * 9.2 = 5.2 \text{ lb/gal} \qquad (4\text{-}5)$$

Thus, pressure reduction at 1000 feet is

$$\Delta P = 14.7\left(\frac{9.2 - 5.2}{5.2}\right) \ln \left(\frac{3.53 * 9.2 * 1000}{1000}\right) \qquad (4\text{-}6)$$

$$\Delta P = 39 \text{ psi}$$

Although the pressure reduction appears to be small, only 39 psi, the resultant mud gradient at 1000 feet is:

$$(9.2 * 1000 * 0.0519) - 39 = 438 \text{ psi}$$

$$438 \div 1000 \div 0.0519 = 8.4 \text{ lb/gal}$$

The mud gradient is reduced from 9.2 lb/gal to 8.4 lb/gal by a reduction 39 psi at 1000 feet. Clearly, with the pore pressure gradient of 9 lb/gal at 1000 feet, the well will kick if this situation is permitted to occur.

***WARNING***

*Gas-cut mud at shallow depths may be extremely hazardous as a severe kick and loss of well control can result!*

These calculations do not take into account the effect of temperature on gas expansion; consequently, the gas volumes calculated at the surface are slightly larger than actual volumes. Hence the amount of mud density reduction is on the high side. Temperature and compressibility have a small effect on gas expansion when compared to pressure. Due to the difficulty of estimating formation temperatures while drilling and of obtaining realistic values for gas compressibility, the calculations above only take pressure into account; the accuracy obtained is sufficient for this particular application. At shallow depths, temperature effect is insignificant and the calculated values are very close to actual gas expansion. At great depth, where temperature change to surface conditions is considerable, the calculated values are optimistic; however, as was shown in the first example, gas-cut mud from deep sections causes no great difficulties.

## 4.6 CUTTINGS CHARACTER

During the normal mud-logging process, cuttings are sieved and graded to a size that is assumed to be representative of drilled cuttings. The larger fragments are cavings from the walls of the borehole and play no part in the compilation of a lithological log.

In geopressure evaluation, these cavings play a major role.

The presence of cavings in the sample indicates that the borehole wall is unstable. The most noticeable and usually most predominant cavings are those of clay, shale, or calcareous lithologies. Coal, however, will cave as a matter of course, hence interpretations should not include coal cavings. The amount of cavings in the bulk sample is an indication of the degree of instability of the borehole walls. Simply watching the cuttings traverse the shaker screens will give a reasonable indication of the amount and size of the cavings in relation to the bulk sample. For this reason it is vital that the Pressure Evaluation Geologist not only supervise how the samples are to be taken, but also regularly check the shakers to see whether caving samples are being ignored.

Cavings are produced due to two mechanisms:

- underbalanced drilling
- stress relief

Abrasion of the walls by the drillpipe will also cause the production of cavings, but generally these will not be discernible from cuttings due to their small size.

If the pore pressure is higher than the hydrostatic pressure in the borehole, the hydrostatic pressure differential will cause the pore fluids to move toward the borehole. In impermeable formations, the resultant pressure gradient adjacent to the borehole wall may become so great as to overcome the tensile strength of the rock. When this occurs, the rock fails in tension, and cavings are formed. This process is illustrated in Figure 4-5.

All parts of the earth's crust contain stresses that change with depth, area, lithology, history, etc. Drilling a hole in the ground relieves some stresses other

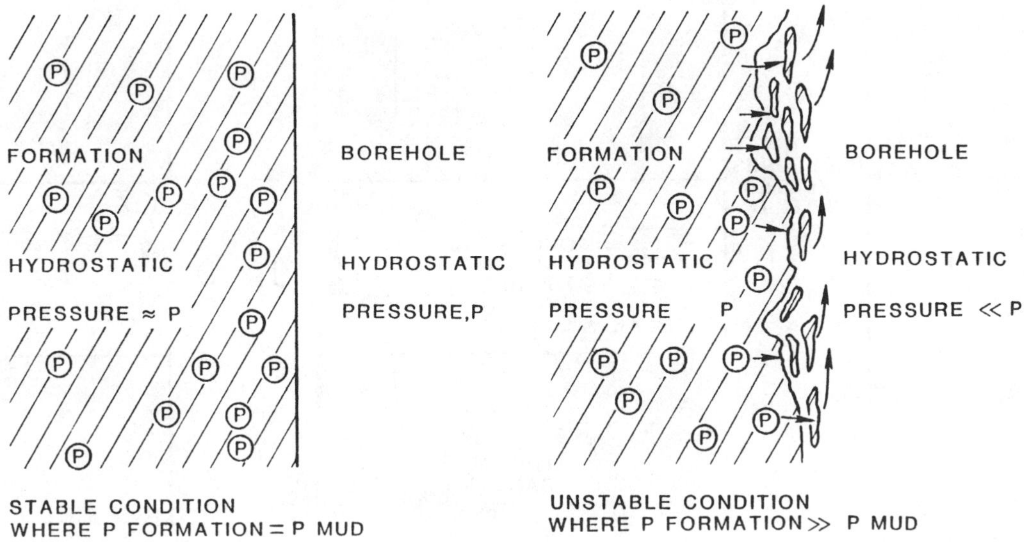


Figure 4-5. Cavings produced due to underbalanced drilling

than those in the vertical plane, and the hole geometry in relation to some stresses acts to concentrate them. If the borehole wall is insufficiently supported by the mud column, it may fail either in compression due to the vertical stress, or in tension due to the horizontal stress, or both. This process is illustrated in Figure 4-6.

The drilling process causes the formation of microcracks and fractures, and these act as areas of stress concentration and potential initial failure points. Thus it is sometimes noticed that part of a borehole may cave copiously for a short period of time, and then become stable. This is due to the removal of the damaged zone (i.e., cavings) adjacent to the bore/formation interface. Formation is exposed which is more coherent and lacks concentrations of stress; thus it absorbs the extra energy without failing.

Cavings produced due to underbalanced drilling are typically long, splintery, concave and delicate. Their typical appearance is illustrated in Figure 4-7a. Cavings produced due to stress relief tend to be more blocky and can vary in size tremendously, depending on the formation characteristics. Examples are shown in Figure 4-7b.

Remember that if the cavings are clays, they may react with the mud and lose their distinctive morphology. Interpretations based on reactive clays should be pursued with caution.

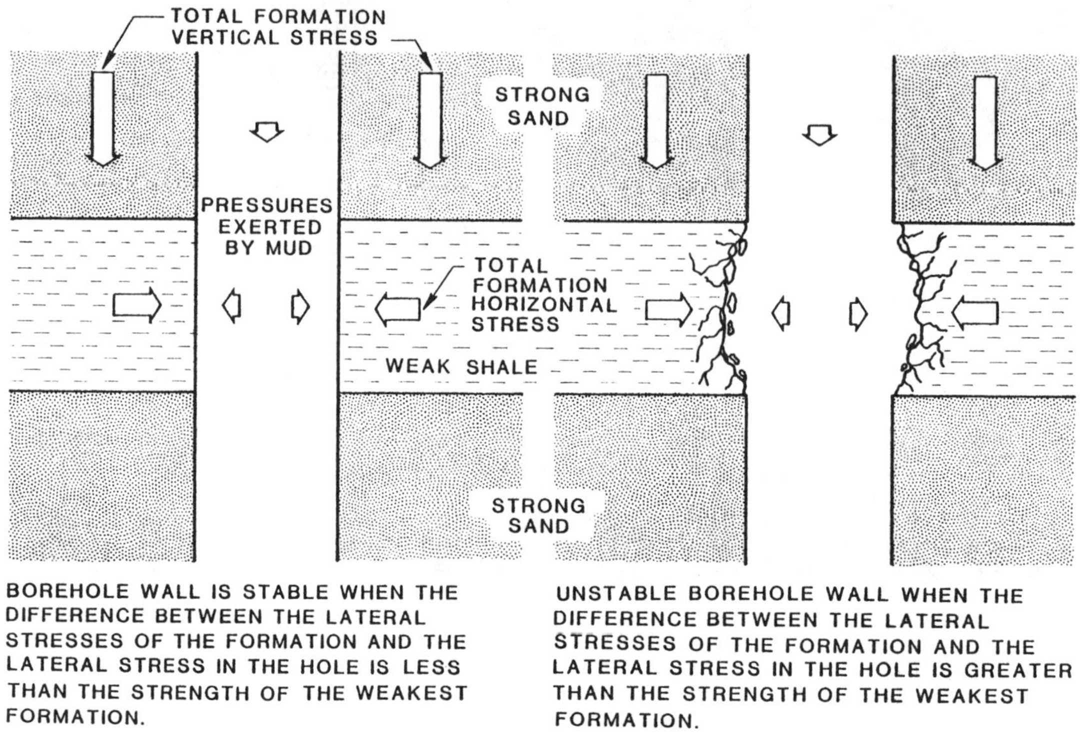


Figure 4-6. Cavings produced due to stress relief and compressional failure

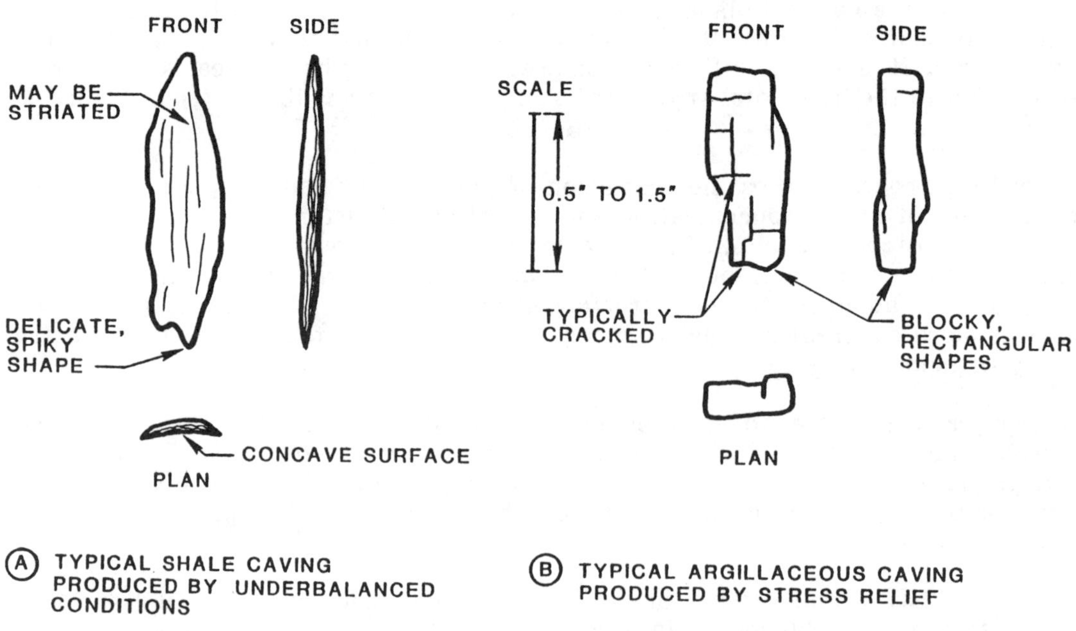


Figure 4-7. Typical cavings produced by underbalance and stress relief

## 4.7 HOLE BEHAVIOR

When a condition of near balance occurs (for example, ECD will balance formation pressure but mudweight alone will not), there is a tendency for fluid to flow into the borehole. If permeability exists, the well may kick. If permeability is low, insufficient fluid will flow to cause a kick, but there will be large shows of trip and connection gases. Where the fluid is unable to flow, spalling or caving occurs. These effects will be recognized by increased torque when drilling, drag on trips and connections, and bottom fill after trips. Normal drag after drilling new hole is of the order of 10,000 to 20,000 pounds, depending on hole and drillstring geometries, but any drag consistently and significantly greater than this is indicative of unstable borehole conditions. Deviated holes will, of course, incur much higher consistent drag.

The occurrence of connection gas indicates that a condition of imbalance exists when the hole is swabbed at connections. Similarly, when localized gas shows (such as trip gas, connection gas, thin sands) do not fall off rapidly but linger, often accompanied by a gradual unexplained increase of background gas, a condition of slight underbalance is indicated.

Clay rocks are a major source of the hydrocarbons normally flushed out of the clays with the pore water during compaction into the permeable zones which constitute the eventual reservoir. Since such flushing does not take place in a geopressured clay rock, the rocks generally carry a far higher hydrocarbon saturation than normal. This will be reflected at surface by an increase in background gas, and, since clay rocks have low permeabilities, by high cuttings gas or blender gas.

This of course is not true of <u>all</u> geopressured clay rocks. If a clay contained no organic debris at deposition, it will contain no hydrocarbons — in either its normal pressured or geopressured state.

Occasionally an apparent paradox may exist: considerable hole drag precludes the possibility of pulling out of the hole, and continued circulation does not release significant debris. An interpretation may be that a degree of differential sticking is occurring, hence to cure the problem the mud density should be reduced. Another interpretation may be that part of the hole is producing cavings that are not immediately circulated out of the hole, in which case the mud density should be increased. Careful analysis of all geopressure evaluation data should indicate whether the problem is due to overbalance or underbalance. If the problem remains unsolvable, the mud density should be first increased slightly to see if the drag is cured; if not, the pipe may become differentially stuck -- but this may be rapidly cured by lowering the mud density to below the original density.

## 4.8 DRILLING EXPONENTS

The rate at which a formation may be drilled is determined by a number of factors, some of which are:

- Force Applied: This is the effective weight-on-bit per unit area of bit cutting structure. This factor includes bit size, tooth shape and distribution, actual weight-on-bit and threshold force (the minimum force at which the bit will drill).

**NOTE**

In areas where the S.I. metric system has become established, it is common to substitute the term force-on-bit for the traditional weight-on-bit. In this manual we will maintain the original term, with the reminder that the terms weight and force-on-bit are in all cases synonymous and refer to the sum of the vertical components of all forces acting upon the bit, the most important of which is the buoyed weight of that portion of the bottomhole assembly which is in tension. The quantity is expressed in units of force, that is pounds-force (lb-f), kilograms-force (kg-f), poundals (pdl), or newtons (N).

- Rotary Speed: The rate at which force is applied and the duration of the force.

- Tooth Efficiency: This is a variable term compounded of the original cutting structure efficiency, minimum effective cutting structure (i.e. the point of tooth wear at which the bit ceases to drill) and the rate at which the bit loses efficiency.

- Differential Pressure: This affects the efficiency of the drilling process by controlling the rate at which cuttings are cleared from bottom.

- Drilling Hydraulics: This is controlled by pump pressure, flowrate, nozzle sizes, and mud rheology. If too little hydraulic action is applied there will be inefficient hole cleaning, and penetration rate will be slowed. Hydraulic action in excess of that necessary for efficient hole cleaning increases penetration rate by jetting action ahead of the bit (see paragraph 4.22).

- Matrix Strength: Although some of the typical sedimentary rock-forming minerals possess high compressive strengths, the binding forces between each mineral grain may be very weak or nonexistent. Hence, an unconsolidated sand has a much lower matrix strength than a consolidated sand. It is similar with carbonates: pore geometry may be such that the matrix may be either weak or competent. Matrix strength may thus be the converse of "drillability" in the drilling industry.

- Formation Compaction: This is related to matrix strength in that it defines porosity distribution; formation compaction simply increases the ratio of matrix material to pore space. Since it is easier to penetrate a pore rather than solid matrix, compaction may not change the actual matrix strength but will affect drilling response as it increases.

When a bit tooth penetrates a hard formation, it forms a cone of crushed rock immediately beneath the tooth, and cracks form in the rock (Figure 4-8). In plastic formations the material will be gouged rather than crushed, but formation of cracks alone will not make hole. The cuttings must be removed as they are formed. The most effective force for the removal of cuttings is the high-velocity jetting action of the bit.

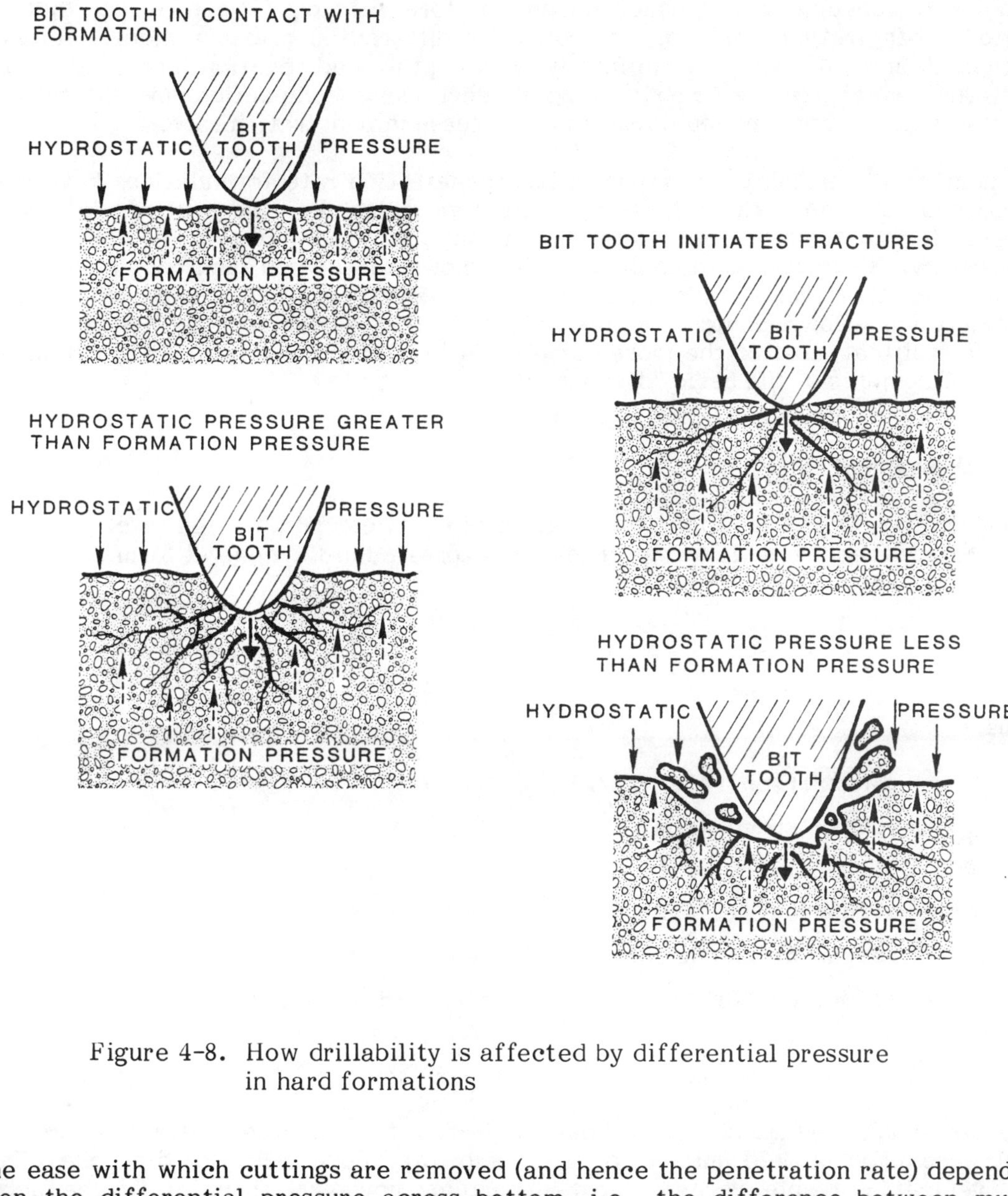


Figure 4-8. How drillability is affected by differential pressure in hard formations

The ease with which cuttings are removed (and hence the penetration rate) depends upon the differential pressure across bottom, i.e., the difference between mud hydrostatic and formation pore pressure. If hydrostatic pressure is in a large excess of formation pressure (overbalance), cuttings will be held down against bottom by the excess pressure differential. As the overbalance is decreased, these effects will be reduced, cuttings will be removed more easily and penetration rate will increase. If formation pressure increases sufficiently for it to exceed hydrostatic pressure (underbalance), mud filter cake ceases to form and cuttings are forced away from the formation, with a consequent increase in penetration rate. It has been known for large 'cavings' to be produced, under conditions of very high underbalance, from beneath the bit. Slight tooth impact causes failure, and upon logging the hole, the caliper logs may show remarkably in-gauge hole, even though the volume of these "cavings" was copious during drilling.

Thus with constant drilling conditions in a uniform lithology, it can be seen that the rate of penetration would be controlled by differential pressure alone. Rate of penetration would decrease uniformly with depth as compaction increased. Upon entering a geopressure transition zone, decreasing compaction and differential pressure across bottom would lead to an increase in penetration rate.

A number of "drillability" or normalized penetration rate formulations have been proposed to remove the effects of drilling variables. For the best application of these formulations, direct data monitoring and computation equipment are necessary. However, field application has shown that, when such equipment is not available, the easiest and most reliable method is the "d-exponent." This formulation allows control of the major drilling variables, and has proved so successful that most of the more complex "drillability" formulations are extensions and refinements of the basic "d-exponent."

## 4.9 D-exponent

Bingham (1965) proposed that the relationship between penetration rate, weight on bit, rotary speed, and bit diameter may be expressed in the general form

$$\frac{R}{N} = a\left(\frac{W}{B}\right)^{d}$$

where

R = penetration rate (ft/min)

N = rotary speed (rpm)

B = bit diameter (ft)

W = weight on bit (lb)

a = matrix strength constant (dimensionless)

d = formation "drillability" exponent (dimensionless)

Jorden and Shirley (1966) solved Equation (4-6) for d, inserted constants in order to allow common oilfield units to be used, and to produce values of d-exponent in a convenient workable range. Most important, however, they let "a" be unity, removing the need to derive empirical matrix strength constants, but making d-exponent lithology specific:

$$d = \frac{\log\left(\frac{R}{60N}\right)}{\log\left(\frac{12W}{10^{6}B}\right)} \tag{4-7}$$

where

d = drilling exponent (dimensionless)

R = rate of penetration (ft/hr)

N = rotary speed (rpm)

W = weight on bit (lb)

B = bit diameter (inches)

In a constant lithology, d-exponent will increase as the depth, compaction and differential pressure across bottom increase. Upon penetration of a geopressured zone, compaction and differential pressure will decrease and be reflected by a decrease in d-exponent (Figure 4-9).

Differential pressure is dependent upon mud density as well as formation pore pressure. Therefore, any change in the mud density used promotes an unwanted change in d-exponent.

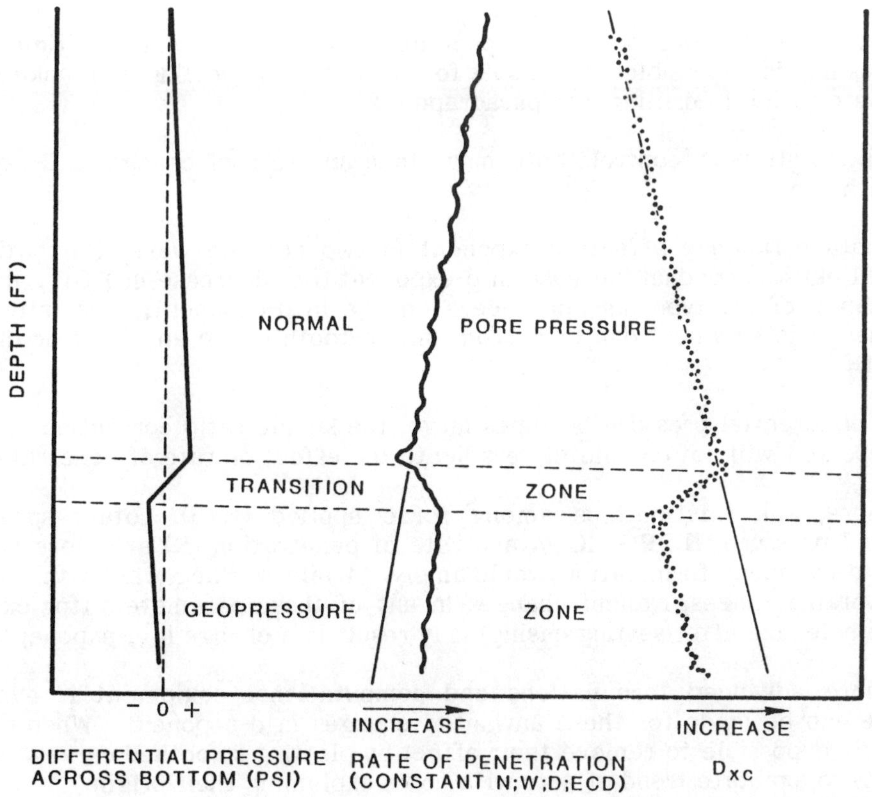


Figure 4-9. Highly stylized curves showing typical response in transition and geopressured zones

Rehm and McClendon (1971) proposed the correction

$$Dxc = d * \frac{N.FBG}{ECD} \tag{4-8}$$

where

d = d-exponent

Dxc = corrected d-exponent

N.FBG = normal formation balance gradient - EMD (lb/gal)

ECD = effective circulating density (lb/gal)

This correction was empirically derived but has been applied worldwide with much success. The use of actual mud density in place of effective circulating density has been found to be acceptable within normal limits of accuracy. ECD should, however, be used when available..

Factors not considered by d-exponent in its basic form are drilling hydraulics, tooth efficiency and matrix strength:

- Drilling hydraulics become important in large holes where efficient hole cleaning is impossible, and in soft formations where jetting will make a large contribution to drilling (see paragraph 4.21).
- Matrix strength controls both magnitude and rate of change of d-exponent with depth.
- Tooth efficiency affects d-exponent in two possible ways: (1) tooth wear will cause a gradual increase in d-exponent (i.e. decrease in ROP), and (2) a change of bit type may produce a change in d-exponent, especially if the change is a radical one (e.g. from milled-tooth bit to an insert or diamond bit).
- If differential pressure becomes large, the simple ratio correction to the d-exponent will not completely eradicate the effect on rate of penetration.

Furthermore, the relationships among force applied (W/B), rotary speed (N), differential pressure (N.FBG/ECD), and rate of penetration (R) are more complex than the d-exponent formulation would imply. While working well within certain normal working ranges, radical changes in any of these parameters (for example, change in hole size after setting casing) may result in a change in d-exponent.

Where more advanced formulations and computational equipment is available, allowance can be made for these unwanted changes in d-exponent. When plotting manually it is possible to remove their effect by plotting smoothed curves. Better practice is to annotate trend offsets with notes explaining their origin.

The d-exponent may be plotted on either semilog or rectangular coordinate grid, and in either case will produce an approximately linear, normal, compaction trend line (see below). Practice has shown that the semilogarithmic grid gives a more efficient data display and is a more suitable format when formation pressure estimates are made from d-exponent.

$$Dxc = \frac{\log\left(\frac{R}{60N}\right)}{\log\left(\frac{12W}{10^3 B}\right)} * \frac{N.FBG}{ECD} \qquad (4\text{-}9)$$

where

Dxc = corrected d-exponent (dimensionless)

R = rate of penetration (ft/hr)

N = rotary speed (rpm)

B = hole diameter (inches)

N.FBG = normal formation balance gradient (lb/gal)

ECD = effective circulating density (lb/gal)

W = weight on bit, 1000 lb

or in the metric form:

$$Dxc = \frac{\log\left(\frac{R}{18.29N}\right)}{\log\left(\frac{W}{14.88B}\right)} * \frac{N.FBG}{ECD} \qquad (4\text{-}10)$$

with R in m/hr

N in rpm

W in tonnes (1000 Kg)

B in cm

N.FBG and ECD in g/cc

A d-exponent plot should be commenced as soon as drilling begins, and ideally should be calculated and plotted every 5 to 10 feet. If penetration rates are high it may be necessary to work in 20-ft intervals or more.

The major causes of "scatter" in a d-exponent plot are:

- Lithological variation: The d-exponent value is dependent upon matrix strength and will therefore change where there are lithological changes. Where lithological variations are relatively minor, e.g. silty laminations in claystone, it is necessary to adjust the normal trend line in order to account for these changes. Where there are major lithological variations (e.g. interbedded sands and shales), it may be necessary to develop a normal compaction trend line for each lithology (Figure 4-10).

- Drilling hydraulics: Where drilling hydraulics are changed, or there is a change in the susceptibility of the formation to jetting, there will be a change in d-exponent. It may be found that shallow unconsolidated sediments will be jetted rather than drilled (see paragraph 4.21). Pump pressure is plotted alongside the Dxc in shallow formations, allowing any fluctuations in pump pressure to be related to the change in Dxc.

- Bit types: Different drilling mechanisms with different bits cause changes in drilling response which is reflected in Dxc scatter and trend offsets.

Offsets due to bit wear should be disregarded (after careful evaluation), and end-to-end curves should not be plotted. Modern, high-speed, soft formation journal bearing insert bits drill just as fast and last longer than comparable milled tooth bits. A past convention, when insert type bits were used, was to subtract 1 inch off the diameter of the bit, in order to avoid shifting trend lines. This practice is not necessary, because insert bits now drill just as efficiently as their milled-tooth counterparts. Further, since diamond bits drill by scraping action alone, RPM is directly proportional to rate of penetration. The Dxc model should thus be more applicable to diamond bits than to roller types; so again, <u>the practice of subtracting 1 inch from the bit diameter should be avoided.</u>

Penetration of a transition zone with a dull bit makes evaluation difficult. Changes in rate of penetration will be less marked, and the decrease in d-exponent due to decreased differential pressure may be partially or even totally masked by an increase due to bit wear.

It is essential that d-exponent not be considered in isolation. Any instantaneous decision based on d-exponent should be conditional upon confirmation (after the lag time) (1) by other parameters, and (2) that there has been no lithology change. Therefore, it is not normally sufficient to trip a bit on the basis of d-exponent alone. Returns should be circulated whenever a d-exponent deviation is seen and before any trip. For example:

- A transition zone is drilled with a dull bit
- No decrease in d-exponent is seen
- A geopressured sand is penetrated with a pore pressure which is just balanced by the mud density

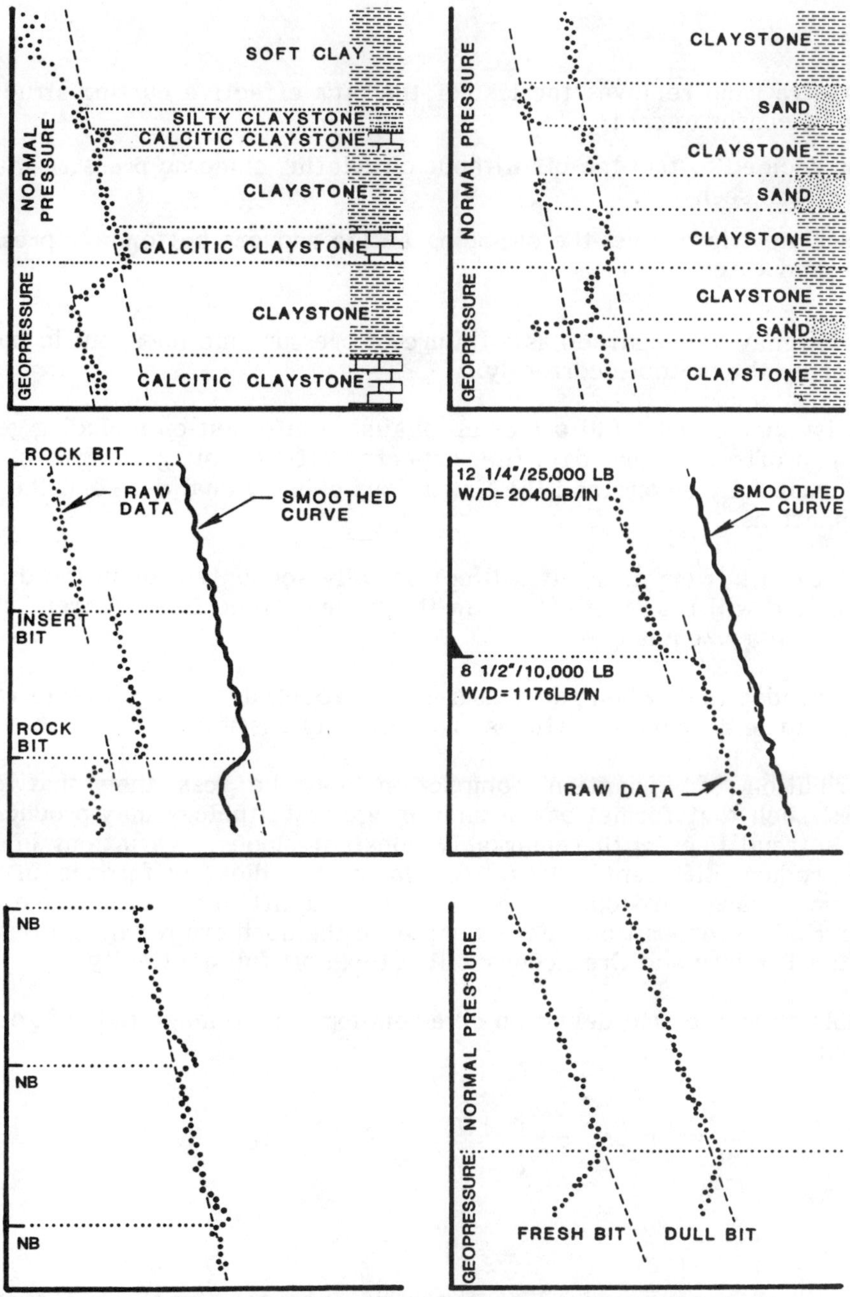


Figure 4-10. Schematic Dxc responses

then,

- Abrasive sand removes the last of the bit's effective cutting structure and the bit ceases to drill
- It is decided to trip the bit without circulating since no pressure indications have been seen
- When the trip begins, the swabbing action reduces bottomhole pressure and the sand kicks

This situation may be construed as a failure of pressure methods, but in fact it is a failure to apply the methods correctly.

The geologist should make full use of all available information including geological prognoses and offset drilling data (i.e. expected bit life, bit gradings when pulled, etc.), and must fully understand the limitations of individual data and the value of data combinations.

Large variations in weight-on-bit will not be fully accounted for in the d-exponent formulation and will result in offsets in the normal trend line. A trend shift may also occur at hole size changes.

It is recommended that, when geopressures are expected, drilling parameters (W, N, B, ECD) should be changed as little as is practicably possible.

The contribution of formation compaction may be less than that of other parameters, such that formations of similar age and lithology may produce normal compaction trend lines with remarkably constant slopes. Variations in lithology may also produce different slopes. Similarly, a radical difference in age may produce some change in slope, especially where uplift and erosion have occurred between periods of deposition. For example, in the northern North Sea Basin, shale trends in the Tertiary and Cretaceous will not exhibit full continuity.

It is possible to relate Dxc deviation on a semilog plot to magnitude of geopressure in the form:

$$P_o = P_n * \frac{Dxc_n}{Dxc_o} \qquad (4\text{-}11)$$

where

$P_o$ = actual pore pressure at depth of interest (psi) or formation balance gradient (lb/gal EMD)

$P_n$ = normal pore pressure (psi) or formation balance gradient (lb/gal EMD)

$Dxc_o$ = observed corrected d-exponent at depth of interest

$Dxc_n$ = expected corrected d-exponent on normal trend line at depth of interest.

Using this equation in the form

$$Dxc_o = Dxc_n * \frac{P_n}{P_o} \qquad (4\text{-}12)$$

with known values of $Dxc_n$ and $P_n$ at some depth, it is possible to substitute values of $P_o$ and calculate the equivalent $Dxc_o$.

Using these calculated values it is possible to plot formation balance gradient lines onto the d-exponent plot parallel to the normal trend line (Figure 4-11).

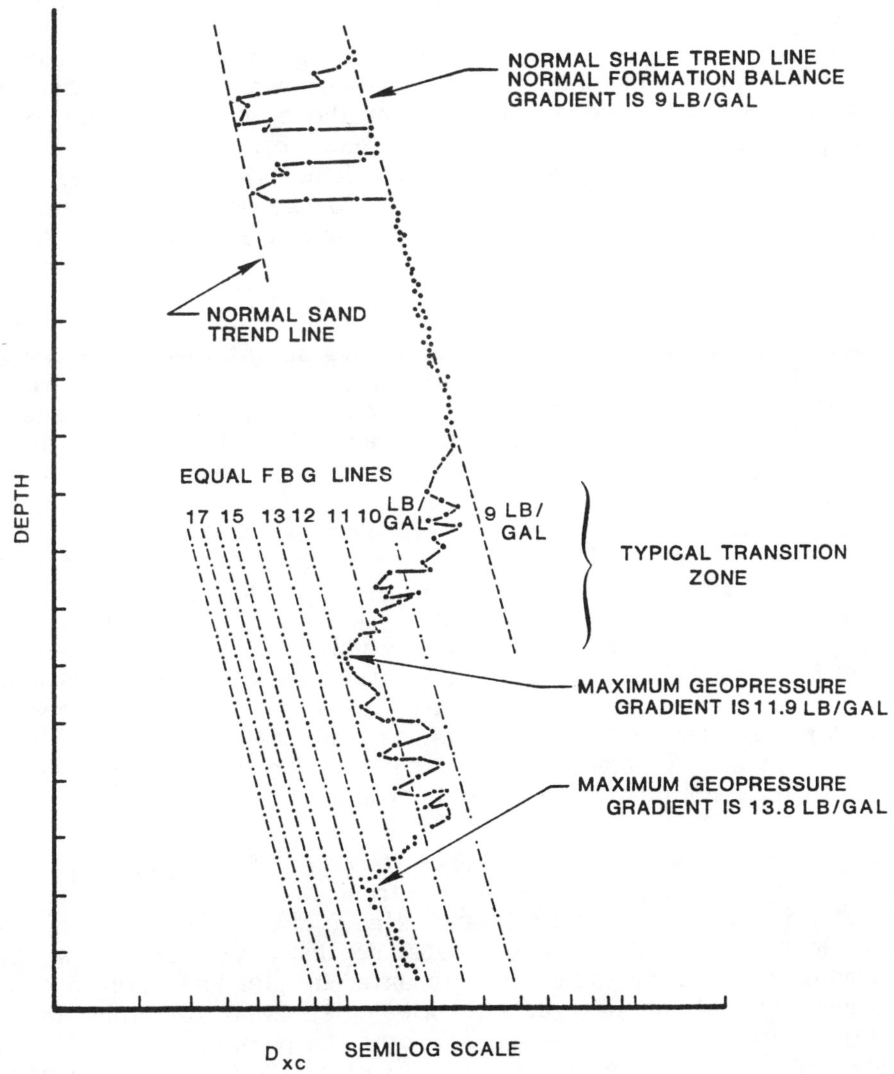


Figure 4-11. Example of formation pore pressure gradients from the Dxc plot

### *CAUTION*

> *Certain transparent overlays are available, ready-marked with equal formation balance gradient lines so that formation balance gradient may be read directly from the log. Such overlays are prepared using Equation (4-12) and a certain depth scale and log cycle. Use of a different depth scale or log cycle will alter the slope and the spacing of the equal formation balance gradient lines and render the overlay useless. Because of the possibility of such errors, <u>transparent overlays should never be used.</u>*

The establishment of trends should be achieved as soon as possible; hence, as drilling progresses, it will be necessary to alter predetermined gradients. The position of normal trends should be established with great care; however, personal selection may be in conflict with another's interpretation. Alteration of trends should not detract from the role of the Dxc as a geopressure indicator, but will change its quantitative meaning. It has been known for normal trends to be changed, thus necessitating reinterpretation of the magnitude of the geopressure zone, by displacing the trend to lower values; however, justification must then be found for the apparently highly overcompacted lithologies above the anomaly. In order that maximum credibility is maintained for Dxc interpretations, all other geopressure indicators must support, as far as possible, conclusions drawn from the plot.

Dxc trend lines are normally placed using two different techniques, which may not be apparent to individual geologists. Some geologists interpret a normal trend in shallow formations and then extrapolate this trend to greater depths. Others interpret normal trends for specific intervals only, changing position and slope to coincide with the majority of points in a particular lithology. Both of these methods contain inherent pitfalls, some of which can render pore pressure evaluation meaningless.

It was stated above that the Dxc normal trend is approximately linear. While this is true over short depth intervals, attempting to extend a linear trend over long intervals is not mathematically correct. Doing so assumes that Dxc is an exponential function of depth, when in actuality it is probably closer to a logarithmic function. If this is the case, a normal trend on semilogarithmic paper will produce a curve that gradually steepens with depth. In a normal-pressure area, this curved normal Dxc trend line is almost universally observed. As of this writing, the Dxc and Nx models are currently under scrutiny — and any major development will be contained in a revised edition.

If the above is true, then as normal trends steepen with depth on semilog grid, it will be necessary to change the straight line trend to a line of greater gradient, but an overall "shift" should not be necessary. Hence extrapolation of a normal trend established in shallow formations to greater depths may diverge from the actual normal trend; and if geopressures are encountered, calculated pore pressures will be in excess of actual magnitudes. Geologists who are thus in favor of extrapolating normal trends should be aware of the possibility that their "normal" trend may not be representative at depth. Geologists who change normal trends with lithological variations may inadvertently steepen trends with depth, reflecting the true behavior of the normal trend on semilog paper. Hence these trends may be more accurate, and pore pressure calculations could be more meaningful. The best rule to follow in trend placement is to make the trend fit the data — not some preconceived idea of how the data should behave.

Scatter and normal trend changes aside, as described above, overall placing of a normal trend (for example, during the latter stages of a well) may be largely dependent on the lithologies encountered previously. As shown in Figure 4-10 (a), the normal trend for the claystone passes through the majority of Dxc points, but falls above the silty zone and below the calcitic horizons. Also, the shallow, unconsolidated clays were subject to jetting, resulting in considerably lower Dxc values. Note that upward extrapolation of the normal trend passes well to the right of these points; however, a curved normal trend, as briefly described above, fits this schematic data well. Figure 4-10 (b) illustrates normal trend development in alternating sands and shales. This diagram represents an extreme, and actual Dxc response in such sequences shows usually an increase in scatter, rather than distinct trend development.

These problems with normal trend-placing accentuate the rule that geopressure magnitude should not be based on Dxc calculations alone.

The above points must also be considered in Nx interpretation. As Nx and Nxb are proprietary information, their formulae are not contained in this manual. Moreover, interpretation techniques must also contain the proviso that the normal trend may steepen with depth, but shifting trends should not be necessary; in fact theoretical justification for a shifted trend is not available.

### 4.10 Nx and Nxb

Nx (normalized exponent) is a refined Dxc that attempts to more closely reflect the various drilling/formation interactions. Where Dxc assumes a linear response between RPM and rate of penetration, Nx models the interaction to a non-linear relationship modified by tooth efficiency and an effective RPM term. Also, the contribution that hydraulics makes in the drilling process may also be normalized, resulting in a drilling exponent that changes more as a result of lithological or pore pressure change, rather than fluctuation caused by bit wear or simplistic drilling parameter modeling.

A plot of Dxc and Nx should show the difference in response clearly; however, the same rules of interpretation (as discussed in paragraph 4.9) apply to both. Drilling into a transition zone with a dull bit may be masked by the Dxc plot, but, because of the tooth efficiency term in Nx, the Nx plot will deviate to lower values, and pore pressure quantification may be achieved by the ratio method.

Nxb is a 'baseline' representation of Nx that attempts to show the overall trend of Nx. In GEMDAS units, it is from Nxb that pore pressures are evaluated. The aim of Nxb is to distinguish pore pressure gradient changes from normal lithological and drilling changes.

Further information on Nx and Nxb may be obtained from the local Exlog office. The discussion of these formation evaluation tools is limited here due to their proprietary nature.

## 4.11 SHALE DENSITY

Shale density determination has often proved to be very effective in determining the degree of undercompaction and consequent abnormal pore pressure in shale bodies. Shale density kits are intended for the rapid determination of the density of drilled shale cuttings. Three methods of shale density determination are available:

1. Single-solution shale density kit
2. Multi-solution shale density kit
3. Mercury pump measurement of bulk density

The single-and multi-solution shale density kits work on the same principle. This is the "Archimedes Buoyancy Principle" which states that a liquid exerts an upward force on an immersed body equal to the weight of liquid displaced.

The kits consist either of a variable-density single solution, or sets of liquids of varying densities. By placing a piece of shale in such a liquid, its density can be determined when it neither sinks nor floats through the liquid.

An accurate determination of shale bulk density can be obtained utilizing the mercury pump. This is known as the "Kobe Method." The difference between the reference volume and the sample volume will determine the bulk density.

Shale density determination can be of great value since it provides information on the compaction of the shale. Under normal conditions, shale density should increase with depth. Any deviation from this consistent trend may indicate that geopressures exist. The magnitude of the bulk density change will vary with the type and magnitude of the geopressure. Often, the bulk density will decrease, but in other cases it may remain constant or continue to increase but at a lower rate than the previously established trend (see Figure 4-12).

It is commonly observed that shale density may decrease as much as 0.5 g/cc or more. If this reduction occurs over a significant depth interval, the calculated overburden gradient may reverse. The low density zone may also change in lithologic character. Fissility, plasticity, carbonate content, color change and other differences may or may not be apparent. Measurements based on cuttings from water-based muds usually are too low, simply due to the adsorption characteristics of clays. Likewise, measurements taken from logs can also give false indications (see paragraphs 4.16 and 4.18). Specifically, the FDC log can be affected by a rugose hole, and the shallow depth of investigation may not read beyond the hydrated zone. The result is erroneously low readings, causing excessive calculated porosities. The sonic log will be greatly affected by hydrated clays, resulting in very high transit times, high porosities and too low densities.

Values may be successfully obtained from these logs when water-based muds are used, but caution should be exercised as errors may exist as explained above.

The best densities are those obtained from wells drilled with less reactive muds, such as diesel types. Both actual cutting densities and log densities should be accurate, as the clay should remain in its virgin state.

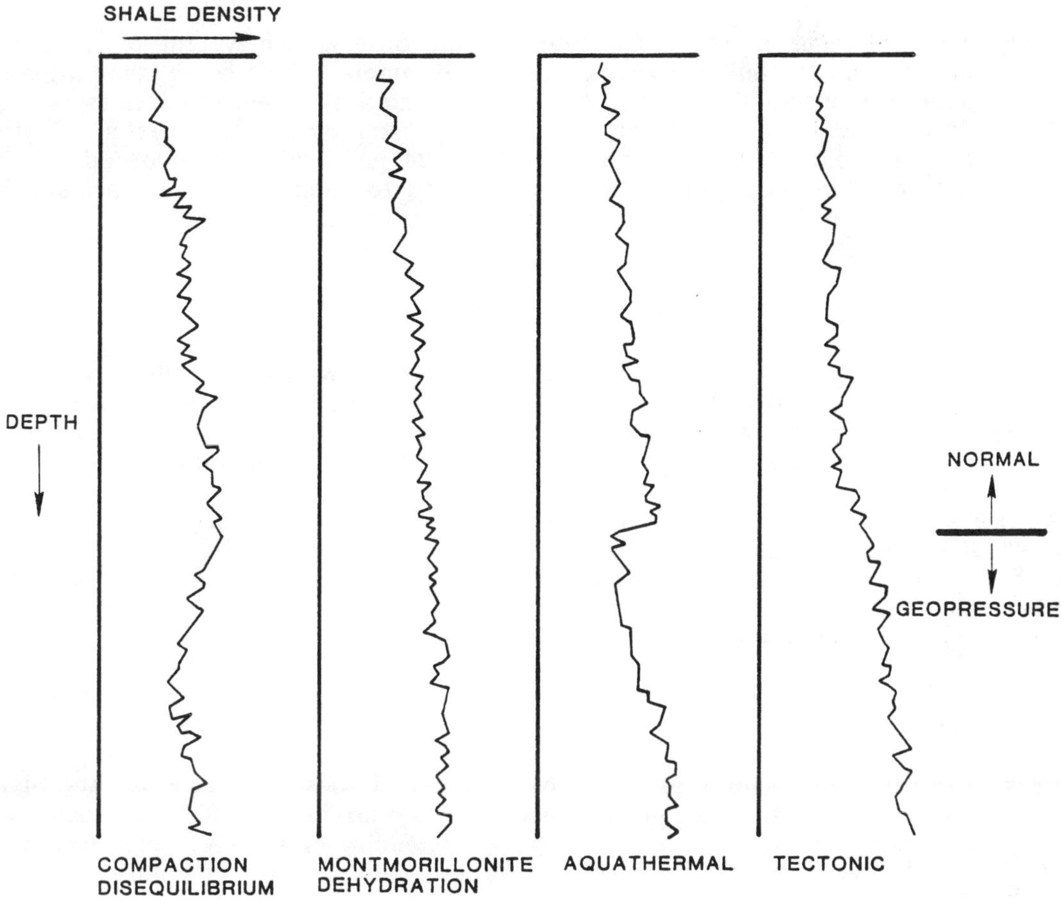


Figure 4-12. Ideal clay density responses in geopressured zones caused by different mechanisms

Several methods are used for measurement of shale bulk density:

- Pycnometer method: Using a container with repeatable volume, this involves measuring change of weight due to displacement of fluid by sample. The most practical application of this method at the wellsite is to use a mud balance.

  Place enough cuttings in the cup so that the balance indicates 8.34 lb/gal (i.e., density of fresh water) with the cap on. Fill the cup with water and weigh again. The new reading is $W_2$ in the following equation:

$$\text{Bulk density (g/cc)} = \frac{8.34}{16.68 - W_2} \qquad (4\text{-}13)$$

- Mercury pump method: The bulk volume of a known weight of sample is measured. The bulk weight of a prepared sample is first established using an accurate chemical balance. The bulk volume of selected cuttings is then determined using a high-pressure mercury pump by the Kobe system (Boyle's Law Principle) at a pressure of about 24 psi, which is recorded on the attached pressure gauge. Mercury is used to compress the air around the cuttings but does not contact the sample material.

**NOTE**

This is contrary to the older procedure in which bulk volume is measured under atmospheric pressure with the bleed-off valve open at the top of the sample chamber, allowing the mercury to contact the sample. This method should not be used.

Accuracy of the instrument and a large amount of sample used (approximately 25 g ≃ 2000 individual shale cuttings) give good consistency of results. Due to the high degree of accuracy and convenience in operation, this method should be used whenever possible; however, very careful and consistent sample handling is necessary for best results.

- Buoyancy method: The sample is weighed in air and in liquid of known density.

- Density comparison methods: The simplest of these is the "Float-and-Sink" method. Shale cuttings are immersed in fluid mixtures of different densities in which they will either float or sink, depending on relative densities. This method is cheap and quick, but is limited in sensitivity due to large difference in densities of available fluids (approximately 0.1 to 0.05 g/cc), and ease of contamination of calibrated fluids.

- Density gradient method: This consists of a fluid column in which density varies uniformly with depth. This is prepared by the partial mixing of a light and a heavy fluid (neothene and tetrabromoethane) in which beads of known density are suspended. A calibration curve of density versus depth is prepared. Shale cuttings immersed in the column will sink to the level at which their density is the same as the fluid. Depth is recorded and density read off from the calibration curve. A major disadvantage of this method is the rapid deterioration of the column due to vibration experienced on some offshore rigs, the expense and time consumption of reproducing the column due to the large volumes and noxious character of the fluids involved; however, with the introduction of the mark 2 version (E/L PN 2302), vibration and fumes should be kept to a minimum.

Both of the heavy liquid methods, while being quick and simple, have the disadvantage of determining the density of individual cuttings. Special care must be taken to ensure that cuttings are true bottomhole cuttings, and several determinations should be made for each interval in order to avoid anomalous results. Six or eight cuttings should be chosen which are representative and free of dust or cracks which may trap air, and of water film which will cause enough surface tension between the water and density fluid to cause erroneous readings.

Increases in density beyond the normal trend due to decreased porosity or calcification should be carefully noted as these may constitute cap rocks above geopressures. Precipitation of pyrite or high iron concentrations result in abnormally high bulk densities in clays and shales. In some wells it has been postulated that the occurrence of pyrite in shales masked the density reduction caused by porosity increase. Careful microscopic examination of clays may indicate the occurrence of very fine pyrite, and high iron concentrations should be indicated by a red/brown color cast. Pore pressure interpretations cannot be accomplished utilizing shale density if heavy minerals are present; however, since shale density is mainly used for qualitative purposes in geopressure evaluation, the role of the other geopressure indicators remains unchanged.

Any decrease in density (without change in clay character) may be recognized as a pressure transition zone.

Recognition of a normal bulk density trend line may be difficult due to a degree of scatter in the rectangular coordinate plot. A semilog plot considerably reduces this scatter, but the normal bulk density range (approximately 1.6 to 2.7 g/cc) results in a more distorted trend line and difficulty in recognizing deviations.

## 4.12 SHALE FACTOR

In paragraph 2.4 it was shown that various clay types have different cation exchange capacities and consequently different adsorption capacities. Also, it was shown that a smectite-rich clay will undergo diagenesis to illite with increasing temperature and ionic exchange. In order that diagenesis may proceed, water must be flushed from the clays. If exchange cations are not available, i.e. potassium, a montmorillonite clay will lose its water but will not convert to illite. Thus if this type of clay is drilled with a water-based mud, the clay will immediately hydrate and cause severe drilling problems.

Shale factor is a measure of the cation exchange capacity of clays. Cation exchange capacity will decrease as clays convert from montmorillonite-rich to illite-rich with temperature (and thus with depth). Pure montmorillonite clays have a C.E.C. of approximately 100 meq/100 g. Pure illites show no swelling characteristics, but their C.E.C. is generally between 10 and 40 meq/100 g. Kaolinites have a C.E.C. of approximately 10 meq/100 g. Of the most common clay types, it is only the smectite group (including montmorillonite) that has an affinity for water. Thus any clay zone that contains montmorillonite will have an affinity for water in an amount proportional to the montmorillonite content, and this will be shown by a proportional value of shale factor. Note that the shale factor as measured at the wellsite will not give values corresponding to actual chemical cation exchange capacity. This is due to impurities in the sample, methodology, experimental error, and the fact that the methylene blue dye (used in the titration) is a very large molecule and thus cannot be adsorbed in interlayer sites.

A reasonably fast method for shale factor determination is described below:

1. Take representative clay/shale cuttings from the sample after it has been dried in the oven.

2. Grind the clay to a fine powder with the pestle and mortar.

3. Weigh approximately 0.5 g of the powder on a balance, and add this to a solution of distilled water and a few drops of 5N sulfuric acid in the metal blender measuring cup.

4. Heat the clay suspension to boiling on the hot plate, stirring continuously.

5. Add methylene blue dye slowly, and regularly remove a drop of the solution on the stirrer and place the drop on filter paper, noting whether the fluid is colored. Generally, the solids in the droplet remain in a localized spot on the filter paper while the water spreads away from the central spot, so that any coloration in the halo may be readily seen.

6. Add methylene blue until the end-point is reached: this occurs when the halo of the blue dye first occurs.

7. Calculate the shale factor:

$$\text{shale factor} = \frac{100}{\text{sample mass g}} * \text{vol ml} * (\text{normality of methylene blue solution}) \qquad (4\text{-}14)$$

where

vol = volume of methylene blue used when end-point was reached

For example:

| | |
|---|---|
| sample mass | = 0.5 g |
| volume of titrate | = 25 ml |
| normality of dye | = 0.01 |

$$\text{shale factor} = \frac{100}{0.5} * 25 * 0.01$$

$$= 50 \text{ meq/100 g}$$

A possibly more accurate but time-consuming procedure has been suggested as follows:

1. Take a clay sample, add about 20 ml of water, and disintegrate sample in the blender.

2. Acidize the suspension with a few drops of 5N sulfuric acid.

3. Sieve the solution through a 180-mesh screen in order to remove sand, lime, etc.

4. Put the suspension into the mud filter press, and allow the water to almost cease flowing from the press before disconnecting the pressured air supply.

5. Weigh 0.5 g of the filter cake that formed on the filter paper.

6. Proceed with the titration in the same manner as described above.

This latter method may be more accurate in gumbo clays.

If the clay is calcareous, and calcimetries are also being run, shale factor may be corrected for carbonate content.

$$\text{true shale factor} = \frac{100}{100 - \text{carbonate \%}} * (\text{apparent shale factor}) \tag{4-15}$$

For example:

A calcareous clay has a carbonate content of 37%, and an apparent shale factor of 16:

$$\text{true shale factor} = \frac{100}{100-37} * 16$$

$$= 25 \text{ meq/100 g}$$

Shale factor can be a useful lithologic indicator, as shown in Figure 4-13. The abrupt shift in shale factor from the normal sand/shale sequences to much more compact sediments at 6705 ft and the break in compaction trend define the top of a 3000-ft section of missing sediments (15 million years) thought to be caused by continental movement and erosion in the shallower continental shelf at that time in the geologic history of Australia. Even in the normally-pressured low-shale-factor shales and carbonates of greater geologic age, there can be occasional anomalies from pressured shale stringers (as at 11,500 ft). See paragraph 4.24 for factors affecting evaluation.

Theoretically, shale factor should be capable of indicating whether montmorillonite dehydration or compaction disequilibrium was the major mechanism in generating an apparent geopressure. As indicated in paragraph 2.2, geopressures caused by compaction disequilibrium indicate that the pressured zone is immature with respect to shallower, normally pressured sediments. This implies that diagenesis has been restricted by the inefficiency of the dewatering mechanism, resulting in clays containing a larger proportion of montmorillonite within the geopressure zone. Shale factor would thus decrease to the top of the geopressured zone, increase within the zone, then decrease as the pore pressure gradients decline (Figure 4-14). Any overall increase in shale factor within a geopressured zone is indicative that compaction disequilibrium has played a part in its formation.

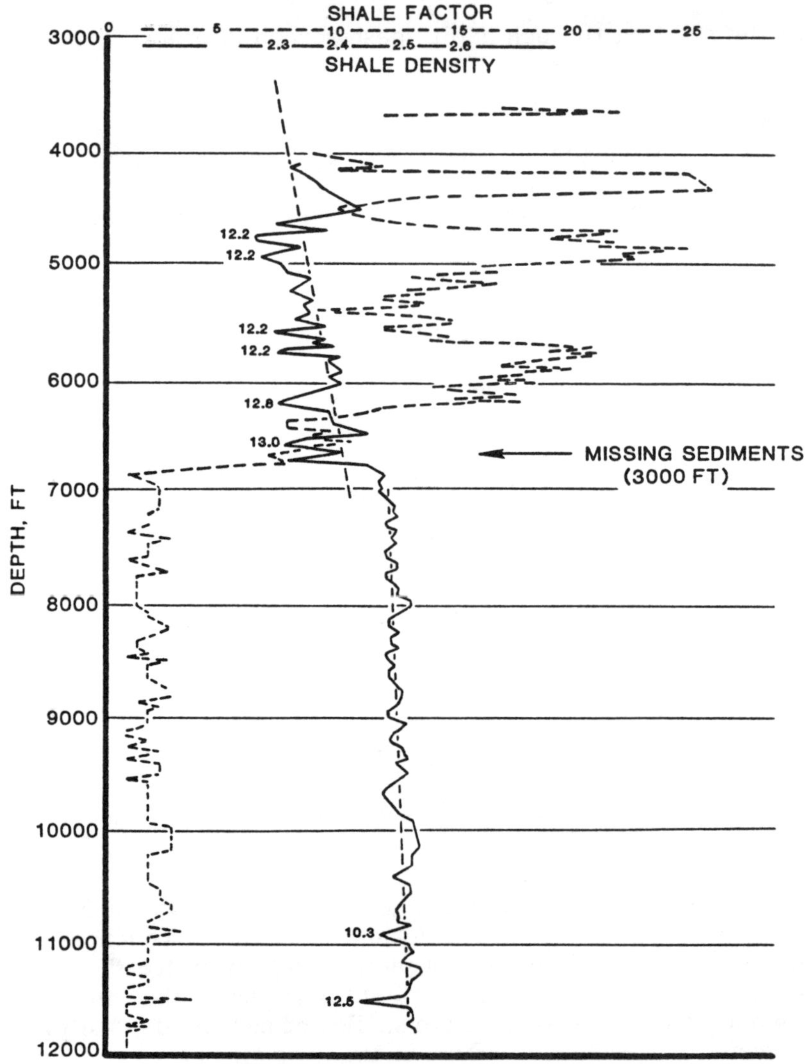


Figure 4-13. Shale factor can be a good indicator of large changes in clay composition, aiding geological interpretation

If, however, a geopressured zone was caused by montmorillonite dehydration (paragraph 2.4), then upon entering the interval, a sharp decrease in montmorillonite content should be observed. Hence the geopressured zone will contain less montmorillonite, as it has been converted to illite — releasing to the pore spaces water which has been unable to escape fast enough and resulting in a pore pressure increase. Shale factor will thus decrease in the pressured zone (Figure 4-14).

Shale factor cannot be a geopressure indicator. The differing responses described above are not definitive, and geopressure has to be indicated from other sources before an interpretation using shale factor can be achieved. Geopressures caused by montmorillonite dehydration <u>and</u> compaction disequilibrium may cause no

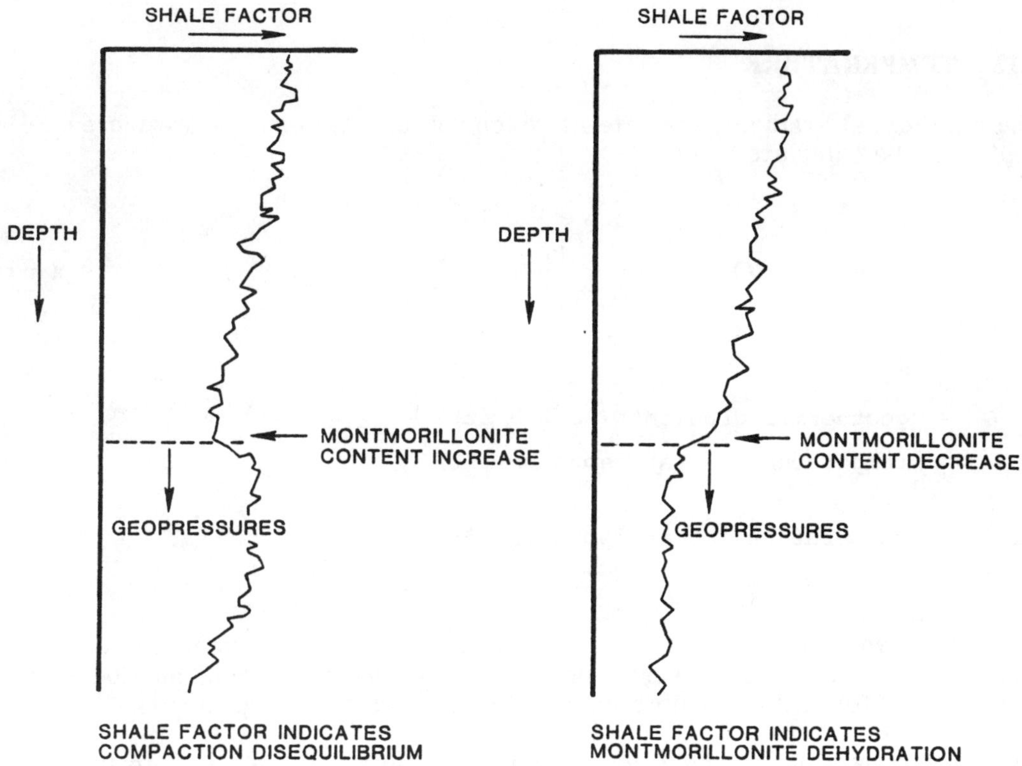


Figure 4-14. Shale factor response in geopressures, caused by compaction disequilibrium or montmorillonite dehydration

change in shale factor; also, if geopressures were caused by another process, i.e. aquathermal pressuring (which is independent of matrix composition), again, a change may not be reflected in shale factor with depth.

In the past, the consensus was that shale factor will increase in geopressured zones and could thus act as an indicator. Reevaluation of the various geopressure mechanisms show that this is not necessarily the case. However, shale factor should be capable of delineating between compaction disequilibrium and montmorillonite dehydration as the major geopressure mechanism.

## 4.13 TEMPERATURE

The geothermal gradient, the rate at which subsurface temperature increases with depth, can be calculated from:

$$G = 100 \frac{\left(T_{F_2} - T_{F_1}\right)}{\left(D_2 - D_1\right)} \tag{4-16}$$

where

$G$ = geothermal gradient (°C/100 ft)

$T_{F_1}$ = temperature (°C at depth $D_1$, ft)

$T_{F_2}$ = temperature (°C at depth $D_2$, ft)

For any given area, the geothermal gradient is usually assumed to be constant. While the average gradient across normally pressured formations may be constant, geopressured formations exhibit abnormally high geothermal gradients.

Since a constant flow of heat occurs radially from the earth's core to the surface, the total flow of heat across any depth increment will be constant. However, the temperature differential across an increment depends upon the thermal conductivity of the material. Since overall heatflow from the earth's surface is generally constant within any particular area, the heat flux through the various formations with depth is in equilibrium. The rate of change of temperature across a formation with a low thermal conductivity (due mainly to high porosity) will be high; conversely, a low geothermal gradient is indicative of high thermal conductivity formations, i.e., lower porosity. Water and hydrocarbon migration to shallower depths may also affect the geothermal gradient. Pore fluids, as insulators, retain heat, so that upon migration these hot fluids modify the temperatures of the formations that they pass through and ultimately become trapped. Note that this mechanism changes the geothermal gradient due to the relocation of hot fluids, rather than attributing gradient fluctuations to porosity. Fowler (1980) cited examples from the Middle East, Canada, and Alaska and other U.S. oilfields, of geothermal gradient bulges which possibly indicated the entrapment of hot fluids from greater depths. The mechanism may also be related to montmorillonite dehydration, in that the huge volumes of water expressed from the clay provide the impetus for migration. "Dead" basins, i.e., no source rocks, have been shown to exhibit normal geothermal gradients, hence on initial exploration wells the geothermal gradient may well indicate the potential of the whole area.

An insulating zone produces a distortion in the isothermal lines which normally run perpendicular to the lines of heat flow (Figure 4-15; Lewis and Rose, 1970). Due to the high geothermal gradient, these are more closely spaced in the insulating zone. In the zones above and below, the isothermal lines are more widely spaced in compensation and the zones exhibit a reduced geothermal gradient. The converse occurs in beds of high thermal conductivity, i.e. sands and some limestones.

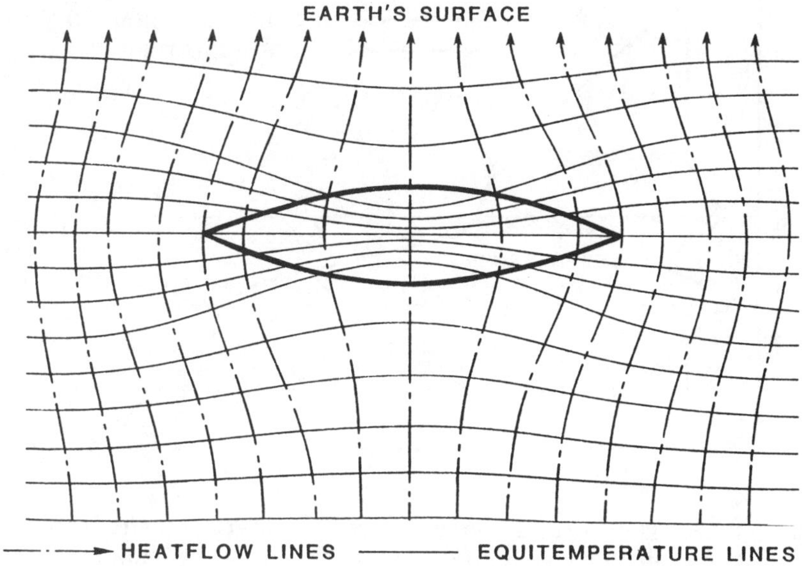


Figure 4-15. Distribution of heatflow and isotherms around an insulating (geopressured) zone

Since water has a thermal conductivity of about one-third to one-sixth that of most rock matrix materials, it can be seen that thermal conductivity is directly related to the degree of compaction of a formation. The higher-than-normal water content of geopressured shales reduces the thermal conductivity. Therefore, the top of a geopressured zone is marked by a sharp increase in geothermal gradient. The temperature of the mud at the flowline may reflect the geotemperature, and recording of flowline temperature is a practical method to determine temperature gradient, provided variable factors such as pump rate, lag time, ambient temperature, lithology, and temperature changes at the surface which are due to mud mixing and chemical treatments can be accounted for. In areas where large annual temperature variations occur, considerable differences may be noted in flowline temperatures: even diurnal temperature fluctuations may cause a 10°C variation in flowline temperature while drilling.

Prior to reaching a geopressured zone, a temperature transition zone will be encountered in which, due to distortion of the isothermal lines, there will be a reduction in geothermal gradient (Figure 4-16). It has been found in practice that this effect is reflected in the flowline temperature gradient, even to the extent of a fall in flowline temperature (i.e. a negative gradient), followed by an extremely large increase in flowline temperature as the geopressured zone is penetrated (Figure 4-17).

A dual temperature probe system with sensors at the flowline and suction pit is effective in removing surface effects, if lagged differential temperature is plotted. It is normally sufficient for the points to be plotted at 30-ft intervals unless more frequent temperature variation is noticed, but points plotted at 10-ft intervals allow more accurate data and better resolution for improved interpretation. Note should be made of circulations, mud additions, water additions, or other significant events.

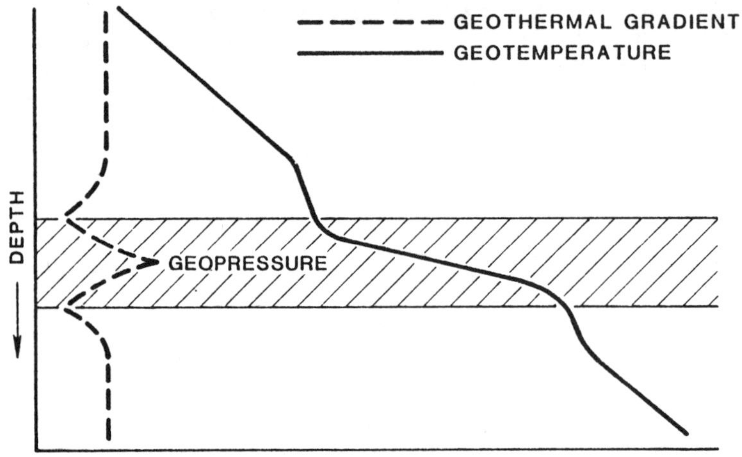


Figure 4-16. Theoretical change of geothermal gradient through an insulating (high porosity/geopressured) zone

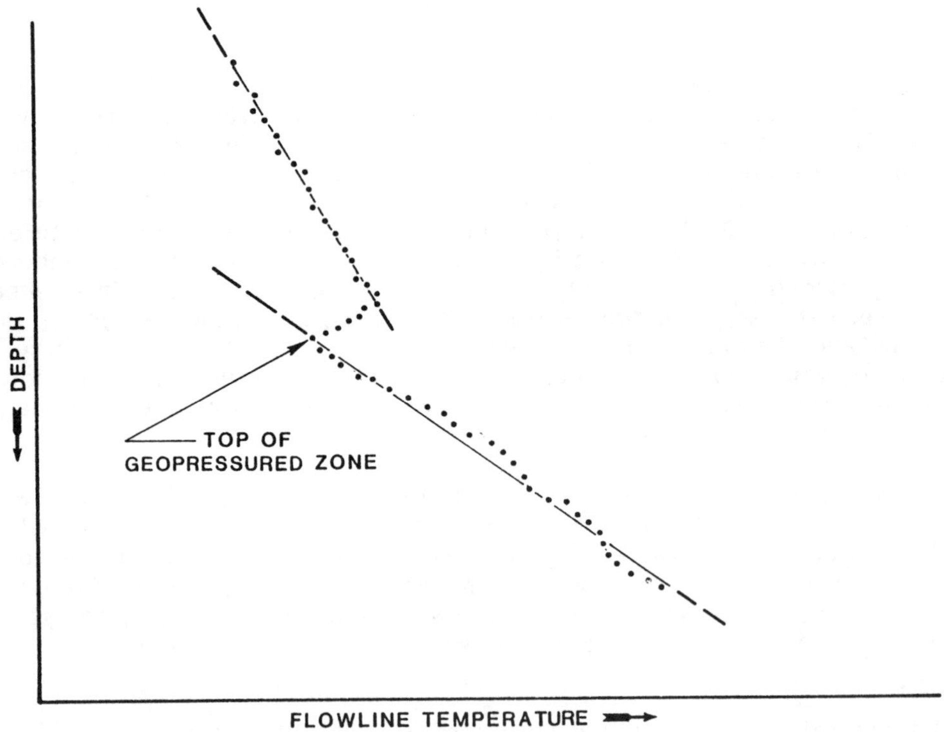


Figure 4-17. Expected flowline temperature response on drilling through a geopressured interval

It is found that the resultant temperature curve is broken when the bit is changed, or during short trips or other downtime, and a certain time is necessary for the mud system to reestablish a temperature equilibrium upon circulation. The rate at which this thermal equilibrium is re-established may be significant, as a more rapid reestablishment may indicate an increased geothermal gradient. Drilling variables which affect the rate of reestablishment of equilibrium include total mud volume. The practice of reducing active pit volume to a minimum, dictated by hole size, aids in reducing the time required to attain equilibrium after tripping and reduces the circulation time needed to stabilize flowline temperature. A discontinuity in the plot also occurs at each casing depth and corresponds to a change in hole size. A higher annular velocity in open hole reduces the amount of heat gained from exposed formations, and a lower annular velocity in the marine riser increases the amount of heat lost to the sea. However, these factors only lead to a change in measured temperature; the rate of change of temperature should remain unchanged. Since pressure predictions can be based on temperature gradient rather than on temperature magnitude, each depth segment between discontinuities can be analyzed separately for gradient trends. It is also helpful to replot a smoothed curve of segments end to end without regard for absolute temperature values. In certain cases it has been found that, instead of plotting the individual segments as an end-to-end smoothed curve, end-to-end plotting of the individual segment trend lines may be of value. This trend-to-trend smoothed curve is merely a graphical method of removing irrelevant scatter from the plot. However, due to a geopressure, the change in flowline temperature may be so small that this curve smoothing may cause the anomaly to disappear. It is therefore suggested that both plots be prepared in order to facilitate interpretations (Figure 4-18).

The reduction in temperature gradient caused by distortion of isothermal lines may be noticed before the geopressured zone is encountered; that is, an advance warning of geopressure may be given. Thus a fall in flowline temperature gradient followed by a sharp rise when the geopressure transition zone is drilled provides a warning that even closer attention must be paid to other drilling parameters in order to achieve confirmation of possible geopressures. However, like other methods of pressure evaluation, flowline temperature reflects a varying physical parameter in an assumed constant rock type; therefore, changes in lithology must be closely monitored in order to avoid false indications.

Wilson and Bush (1973) proposed that flowline temperature gradients can predict geopressure occurrence by use of a gradient factor:

$$GF = \frac{G}{Gn} \tag{4-17}$$

where

GF = gradient factor

G = flowline temperature gradient

Gn = normal geothermal gradient

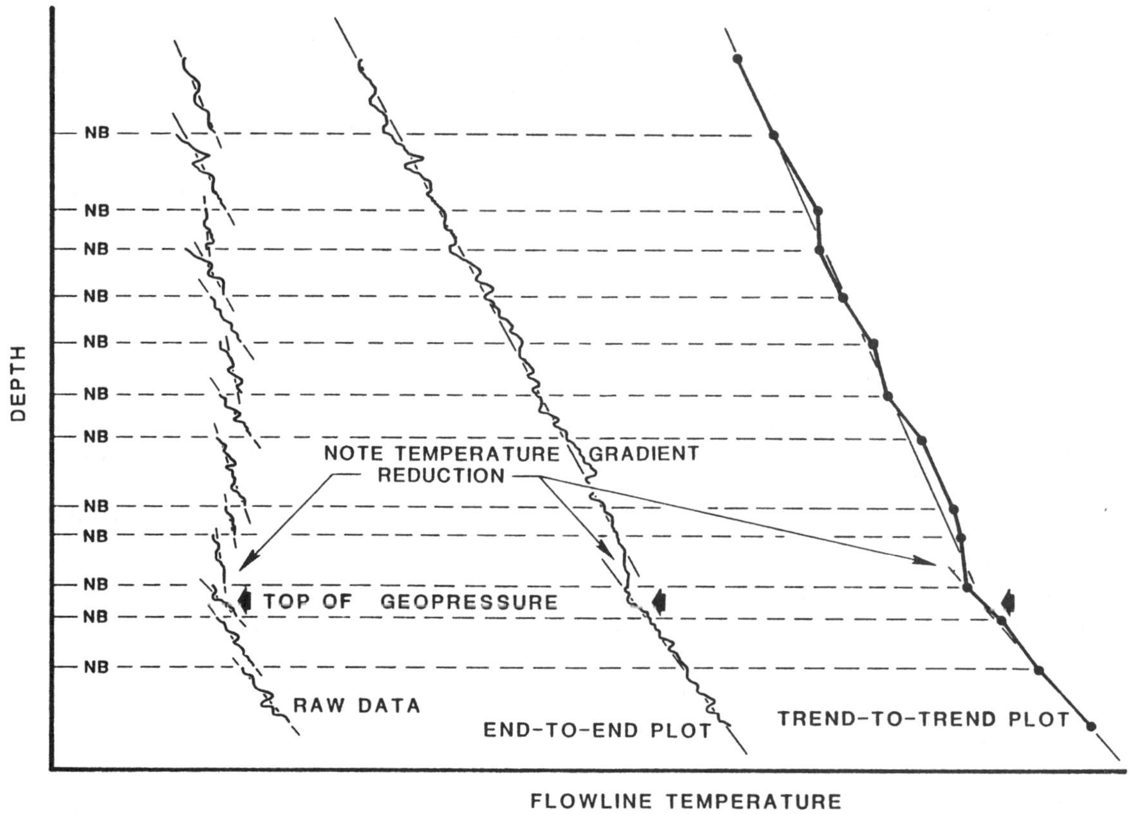


Figure 4-18. Plots of flowline temperature, smoothed end-to-end plot and trend-to-trend plot

The gradient factor may be calculated for each 100-ft interval, then averaged every 100 feet for the preceding 200-ft interval. Zero and negative temperature gradients are recorded as zero values. Apparently, a gradient factor of 2.0 or more is indicative of a geopressure. However, due to the unreliability of most flowline temperature plots in reflecting actual geothermal gradients, and the possibility that gradient factor may not be representative within a particular area, this method should be treated with caution. It may be more valuable for onshore wells.

A plot of maximum temperature on regaining circulation after a period of downtime can closely approximate geothermal trends. After a trip, mud temperature will reach a maximum on bottoms-up. Monitoring these peaks may aid geothermal trend interpretations.

Wireline log temperature data and Temp-Plate data may also be plotted, in addition to plotting

- flowline temperature
- end-to-end flowline temperature
- trend-to-trend flowline temperature
- differential mud temperature ($\Delta T$)

Most electric logging tools contain a maximum-recording thermometer. The maximum temperature is recorded on each log heading and is usually seen to increase with time as the logging program progresses. By use of a modified Horner plot, it is possible to estimate true formation temperature. It is assumed that the maximum temperature occurs at total depth (TD). The Horner expression was originally developed for pressure build-up predictions for reservoir analysis, but was modified by Dowdle and Cobb (1975) to model temperature build-up. Although mathematically incorrect, actual formation temperature may be closely estimated, particularly when circulation periods are short. The theory of the calculation is that, during drilling and circulation, the cool mud reduces the temperature of the formation. This results in a temperature gradient that increases away from the borehole, to a point where the formation temperature is undisturbed. When circulation ceases, heat is transferred to the mud in the borehole, the temperature gradient decreases, and the radius of disturbance decreases. Hence by extrapolating the temperature increase to infinite time, it should be possible to calculate the actual formation temperature. This expression is:

$$T = T_f - C \log \frac{t_c + t_L}{t_L} \qquad (4\text{-}18)$$

where

$T_f$ = true formation temperature

$T$ = measured temperature

$C$ = constant

$t_c$ = circulation time at TD

$t_L$ = time since circulation stopped

Thus a plot of T against $(t_c + t_L)/t_L$ on semilog paper should be linear, and when extrapolated to a time ratio of unity, the result should be a close estimate of formation temperature (Figure 4-19).

The same points from Figure 4-19 have been replotted on Figure 4-20, showing a possibly easier method of displaying the data. Note that the grid on Figure 4-19 is semilog, whereas it is linear in Figure 4-20. Points on Figure 4-20 were plotted using the relation

$$\log \frac{t_c + t_L}{t_L} \qquad (4\text{-}19)$$

against measured temperature. Extrapolation of points using the latter relationship allows a smaller margin of error when drawing lines through a scatter of points. The near normal intercept of the gradient with the temperature axis allows precise temperature determinations.

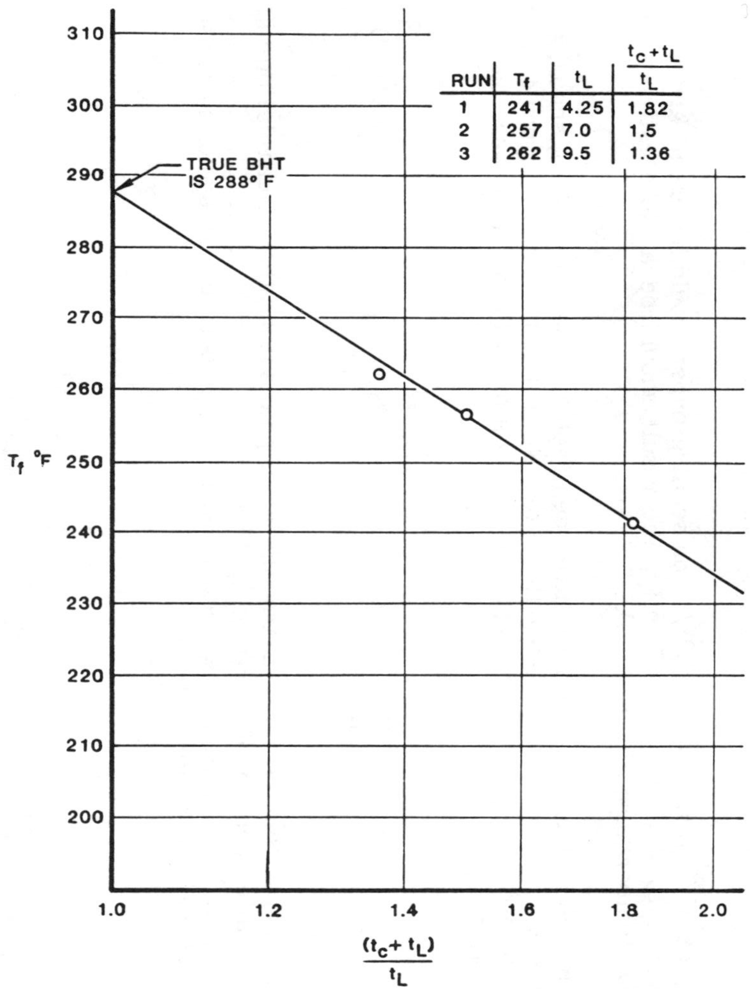


Figure 4-19. Horner-type plot for graphic solution of true bottomhole temperature

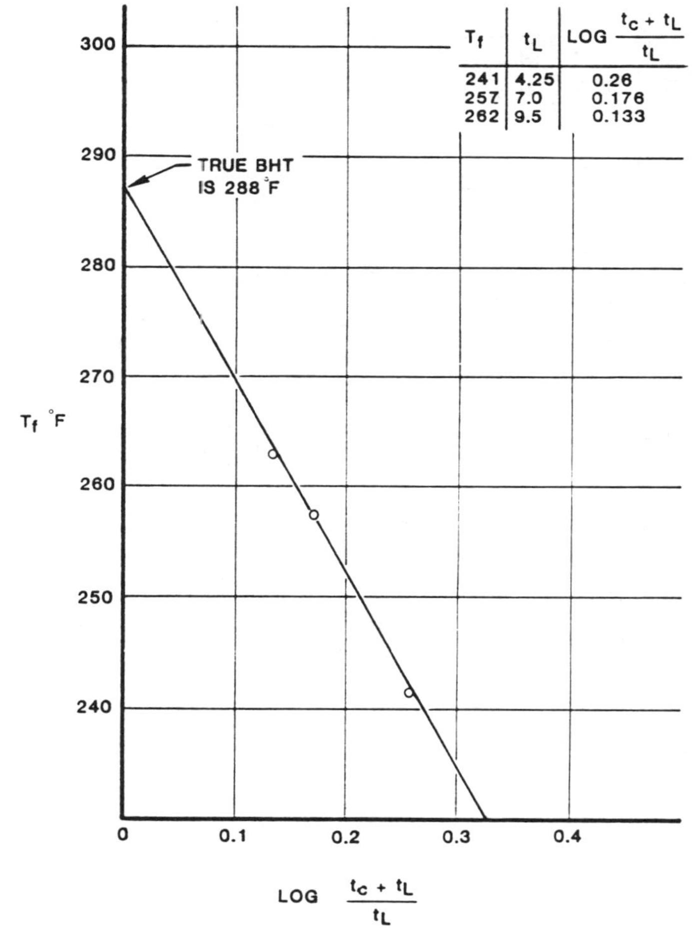


Figure 4-20. Horner plot of linear X-axis. Note less scatter in points

A mathematical method was proposed by Nwachukwu (1976) which utilizes a modified Lachenbruch-Brewer equation. If three temperature points are available, this can prove to be a useful cross-check with the Horner plot. The equation must be solved for $T_f$:

$$\frac{T_f(t_2-t_1) + [(T_1*t_1) - (T_2*t_2)]}{T_2 - T_1} = \frac{T_f(t_3-t_1) + [(T_1*t_1) - (T_3*t_3)]}{T_3 - T_1} \quad (4\text{-}20)$$

where

$T_1$ = recorded BHT, log run 1

$T_2$ = recorded BHT, log run 2

$T_3$ = recorded BHT, log run 3

$t_1$ = time since circulation stopped, log run 1

$t_2$ = time since circulation stopped, log run 2

$t_3$ = time since circulation stopped, log run 3

$T_f$ = true formation temperature

For example,

log run 1, time since circulation stopped = 4 hours

log run 2, time since circulation stopped = 7 hours, 50 minutes

log run 3, time since circulation stopped = 11 hours, 10 minutes

Measured temperatures were 210°F, 219°F and 225°F.

Hence,

$$\frac{T_f(7.83-4) + [(4*210) - (7.83*219)]}{219 - 210} = \frac{T_f(11.16-4) + [(4*210) - (11.16*225)]}{225 - 210}$$

$$\frac{T_f * 3.83 - 874.8}{9} = \frac{T_f * 7.16 - 1671}{15}$$

$$T_f = \frac{(15 * 874.8) - (9 * 1671)}{(15 * 3.83) - (9 * 7.16)}$$

$$T_f = 274.2$$

Depending on the particular environment, one method may be found to be more accurate than the other, but for wildcat use they both should be sufficiently close to actual formation temperature as to make no difference.

At each logging point the estimated bottomhole temperature should be plotted, and between the successive depths the average geothermal gradient may be calculated from Equation (4-16).

A useful geothermal gradient check between log runs can be achieved by use of Temp Plates. These can be more accurate than flowline temperature and $\Delta T$ monitoring, particularly on offshore rigs.

Temp Plates are self-adhesive sensors containing hermetically sealed heat-sensitive elements which change chemical structure at given calibrated temperatures. When exposed to the rated temperature, the indicator turns from pastel grey to black. This chemical reaction is completed in less than 1 second and is accurate to within ±1% of the calibrated temperature. The change is also permanent and irreversible.

Stick the Temp Plate to the survey tool as shown in Figure 4-21, ensuring that the Temp Plate does not come adrift, by wrapping it with tape. It is wise to put a higher range Temp Plate on the clock when the present Temp Plate has two spots exposed.

The Temp Plate should be left on the survey tool until all spots are exposed. The spots may turn light grey with repeated exposure to near reactive temperature, but they will not turn black until the reference temperature has been exceeded.

***CAUTION***

*Do not place the Temp Plates on the exterior of the go-devil, Schlumberger tool, etc. Field testing has shown that contact with diesel muds and high pressures (greater than 2000 psi) render measurements useless. They must be placed in a sealed environment, isolated from pressure and reactive fluids.*

Use caution when evaluating the Temp Plate readings since the condition of the mud system and the plate's position in the drillstring affect its performance. The length of time circulation was terminated prior to running the survey tool affects mud temperature stabilization, and this time period increases with depth. Since steel is a relatively good conductor of heat when compared to mud, high temperatures generated by the rotating drilling assembly and bit may produce artificially high readings — particularly if there has not been sufficient circulation time to dissipate it. Although the Temp Plates turn black upon reaching the reactive temperature, they will pass through darkening shades of grey before reaching this point, but this transition is very rapid.

The adhesive strength of Temp Plates is very good. This means that a wide range of plates can be stuck onto the survey tool early in the operation, and they need be removed only when all the heat sensors on a plate have been exposed. This ensures that minimal interference occurs with the running of the survey and that a successful reading is achieved.

Temp Plates are available in various ranges with a resolution of ±1% of their relative value, and each contains four calibrated temperature indicator disks. Each Temp Plate has a $20^{\circ}C$ range. The total temperature interval including all the Temp Plates is from $35^{\circ}C$ to $215^{\circ}C$ (EL P/Ns 14776 through 14784).

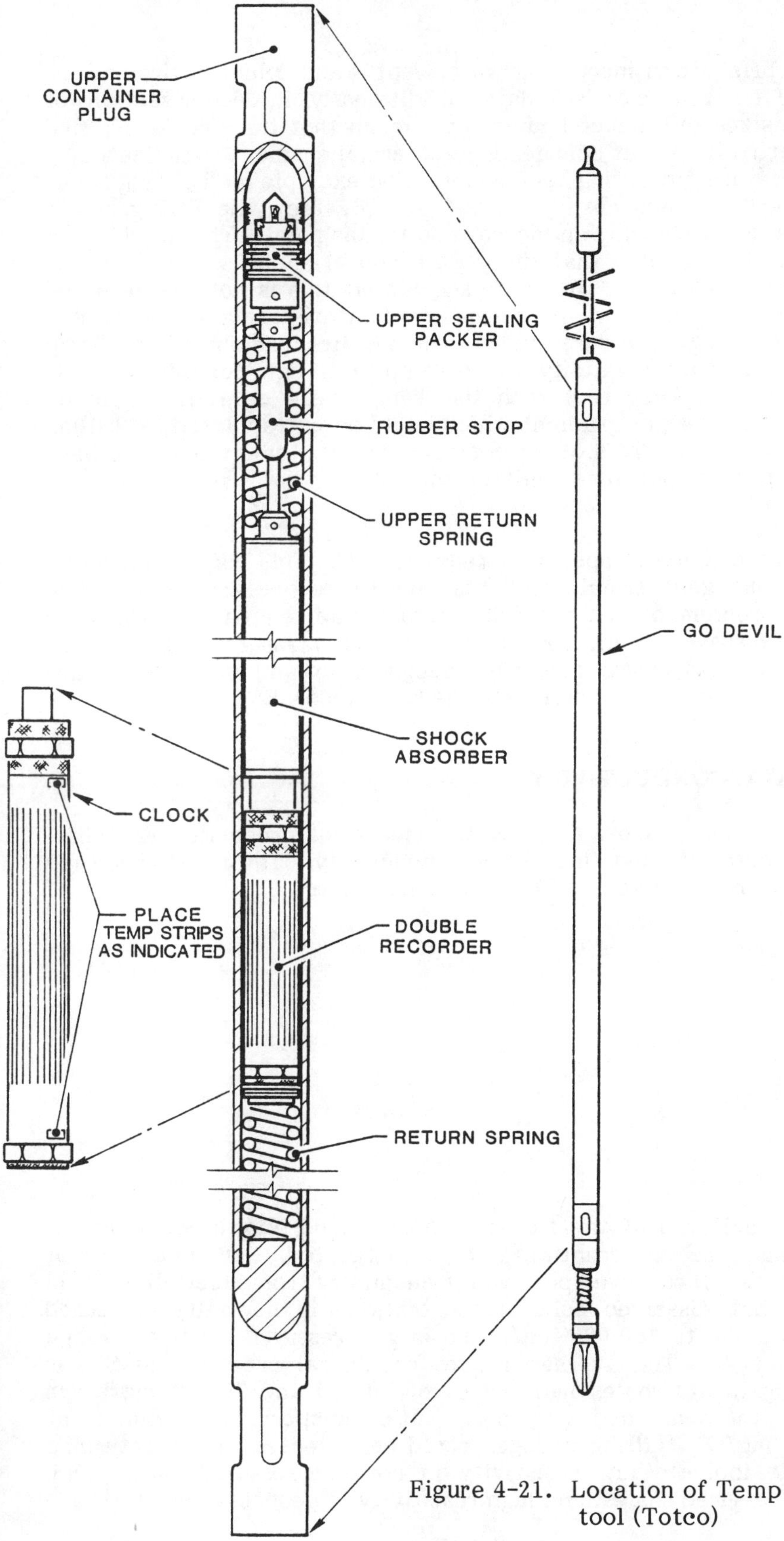


Figure 4-21. Location of Temp Plates in deviation survey tool (Totco)

Today, drilling occurs offshore in increasing water depths and colder environments, resulting in reduced effectiveness of FLT data. Additionally, in deeper sections of some wells, small hole sizes and reduced pump rates mean that the circulating mud at the surface does not truly reflect changes in geothermal gradient as indicated by other downhole temperature measuring techniques. The example well X (Figure 4-22) shows that after a 9-5/8-inch casing was set at 7960 feet, the FLT gradient became almost vertical after showing an increase up to this point. This would have led the logging geologist to assume that the geothermal gradient was also stable. Analysis of both the Temp Plate and BHT data shows that this is not the case and that, in fact, a significant increase in geothermal gradient has occurred — possibly indicating the presence of a geopressured zone. The parallel tracking of the Temp Plate and BHT curves indicate the validity of the Temp Plate data for this well. By correlating the true BHT at 7970 feet with the Temp Plate data, it should be possible to establish a geothermal gradient which could be extrapolated as drilling proceeds. This may be of use where a client company decides that it does not want to drill beyond the upper temperature limit of the "oil window" which is usually considered as being at 170°C.

Note that the resolution of each temperature range is 5°C. This may be too small to detect the anomalous geothermal gradients due to geopressures, and when compounded with the problem of not recording actual mud temperature (because the plate is inside the survey tool), actual temperature measurements may be questionable. However, gross trends should be recognizable, and one of the major advantages of using Temp Plates is to delineate the "oil window."

## 4.14 MUD RESISTIVITY/CONDUCTIVITY

The standard units for mud or formation water electrical properties are ohm-meters for resistivity and millimhos/meter for conductivity. The usual range for muds and formation waters is between 0.01 and 10 ohmmeters.

$$R = \frac{1000}{C} \tag{4-21}$$

where

R = resistivity (Ωm)

C = conductivity (mmho/m)

When conductivity is monitored at the flowline and the mud pits, a conversion is made to chlorides, and the differential, Δ Cl, is purportedly an indicator of geopressures. Schmidt (1973) surveyed pore water chemistry from sidewall cores in Louisiana and found that dissolved solid concentrations in normally pressured sandstones were around 600 to 180,000 mg/l; and in geopressured sands the range was 16,000 to 26,000 mg/l. The average range for normal pressure sands was 120,000 to 170,000 mg/l, and shales had ranges of 10,000 and 70,000 mg/l. In geopressured sections, the shale and sand pore water composition is similar at approximately 20,000 mg/l. If these changes could be detected in the returning mud, the meter would indicate low resistivity in normal-pressured sands, high resistivity in normal-pressured shales, and high resistivity in geopressured sands and shales.

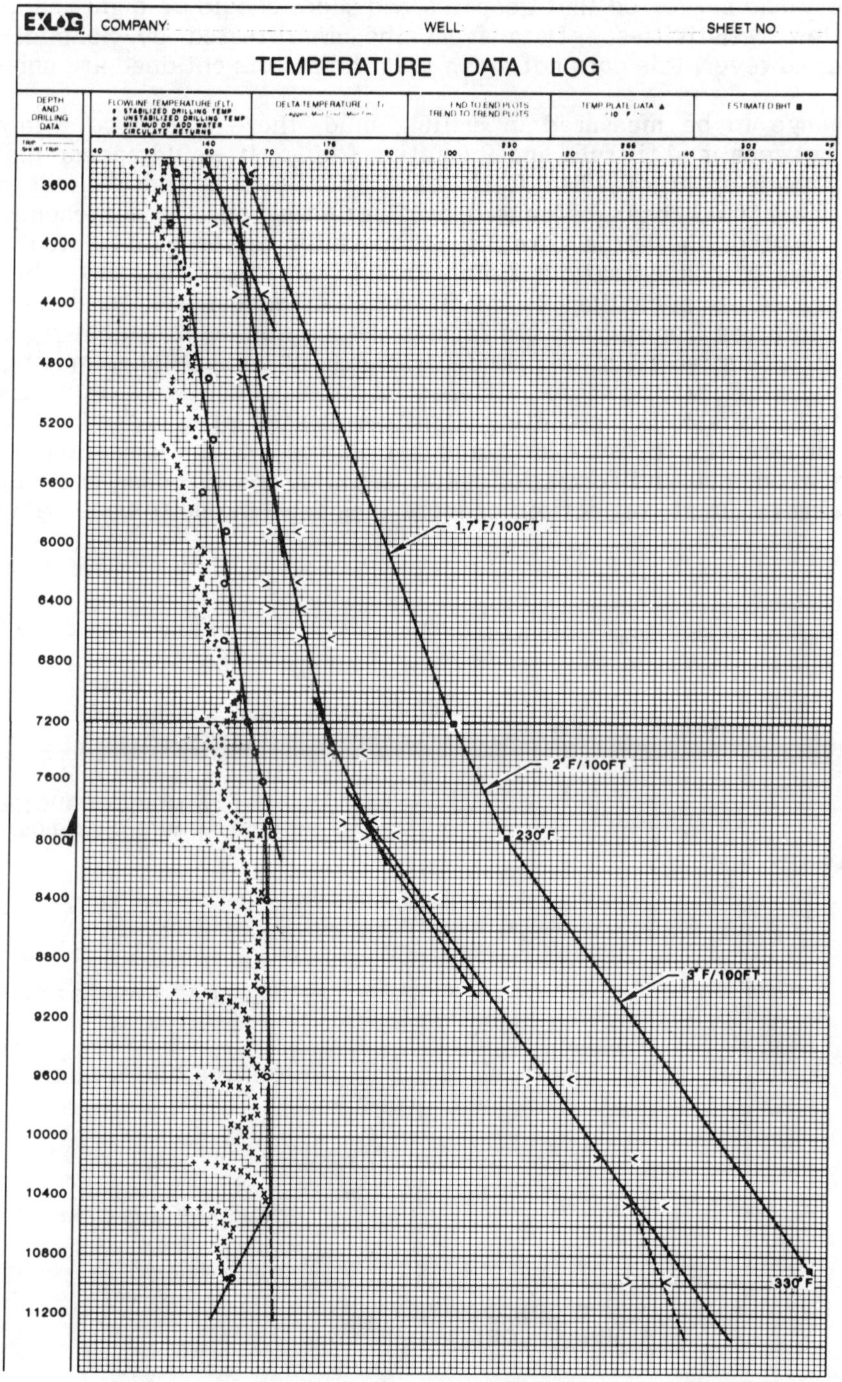


Figure 4-22. Relationship between FLT, Temp Plate, and BHT data for Southeast Asia well

Past theory had suggested that geopressured sands should be highly saline and thus produce low resistivities. Data from the area tested by Schmidt suggested otherwise; however, this does not mean that the results obtained are universal.

For a change to be measured in drilling mud, there should be a large salinity contrast between mud filtrate and formation fluids. It would appear that a change would be more apparent when fresh water muds are in use. Saline muds would severely mask small changes caused by fluctuating pore water chemistry. It is strongly doubted whether a flowline conductivity sensor could detect changes in formation water concentrations simply due to the fact that the volume of pore water released from cuttings is infinitesimal in comparison to the mud volume. However, influxes from permeable formations may be seen as changes either way, depending on relative salinities; hence, warnings of underbalance may be given. In the U.S. Gulf Coast, however, differential mud conductivity or "delta chloride" appears to be reliable in pinpointing slight pore water influxes, and as such is thus a valuable differential pressure indicator (see Figure 4-23). If sufficient difference exists between mud and formation water salinity, the response is similar to ditch gas, showing influx at connections or increasing feed-in due to underbalance.

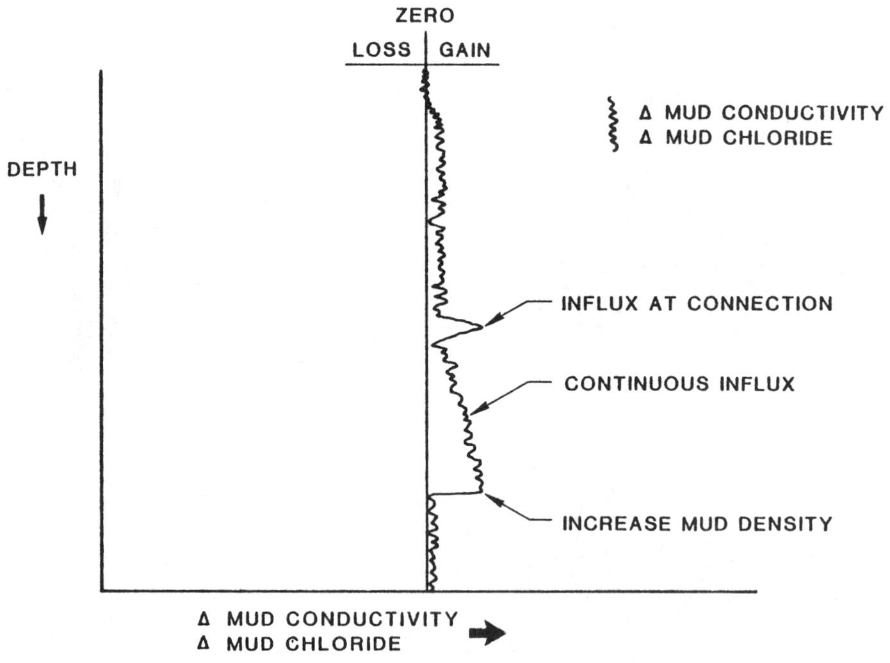


Figure 4-23. Differential mud conductivity and delta chloride log

## 4.15 ELECTRIC LOG PARAMETERS

## 4.16 THE SONIC LOG

The sonic tool measures the interval transit time of the formation. As the distance between the transmitters and receivers is fixed, the only variable is time, hence interval transit time is measured in $\mu$sec/ft.

Compressional waves travel at approximately twice the velocity of other wave types. The reciprocal of velocity, or time in seconds necessary for the compressional wave to travel a unit distance is

$$T_c = \sqrt{\frac{\rho(1+\mu)}{3M_b\ (1-\mu)}} \qquad (4\text{-}22)$$

where

$T_c$ = time

$\rho$ = density of the material

$M_b$ = bulk modulus of elasticity (compression)

$\mu$ = Poisson's Ratio

Hence, acoustic travel time is explicitly dependent upon the density and elasticity of the material. Since different minerals posess different densities and elasticities, laboratory measurements must be undertaken to determine their particular properties. Once these are known, it is seen that the interval transit time for a particular rock will be a measure of its porosity. Porosity may be calculated from:

$$\phi = \frac{\Delta t - \Delta t_m}{\Delta t_f - \Delta t_m} \qquad (4\text{-}23)$$

where

$\phi$ = porosity (fractional)

$\Delta t$ = transit time of particular formation

$\Delta t_f$ = transit time of pore fluids (or filtrate, as the sonic tool only measures approximately 1 inch into borehole wall)

$\Delta t_m$ = transit time of matrix

Note that porosity is usually expressed as percent; however, the value must be fractional when used in any log analysis equations.

Figure 4-24 shows some typical matrix and fluid transit times.

| Formation | $\Delta t_m$, μsec/ft$^{-1}$ |
|---|---|
| Sandstone: | |
| Unconsolidated | 58.8 or more |
| Semi-consolidated | 55.6 |
| Consolidated | 52.6 |
| Limestone | 47.6 |
| Dolomite | 43.5 |
| Clay/Shale | 167-62.5 |
| Anhydrite | 50.0 |
| Gypsum | 52.6 |
| Quartz | 55.6 |
| Salt | 66.7 |
| Granite | 50.0 |
| Iron (casing) | 57.0 |
| **Fluids** | **$\Delta t_f$, μsec/ft$^{-1}$** |
| Salt water | 189 |
| Fresh water | 218 |
| Oil | 238 |
| Methane | 626 |
| Air | 910 |

Figure 4-24. Transit times for matrices and fluids

Since geopressures originate mainly in clays, it can be seen from Figure 4-24 that attempting to calculate porosity could be a problem. The very high transit times apply to the "house of cards" type of structure in montmorillonite clays (Figure 2-13) typical in shallow, wet sediments; the lower transit times are for the more consolidated types. Porosities calculated for clays tend to be slightly high, but corrective factors are not yet available.

A sonic log run in a geopressured clay interval will show increasing transit time, ($\Delta t$), (hence increasing porosity) with increasing pore pressure gradient. In the transition zone (if it exists), the $\Delta t$ curve will be seen to steadily move to the left (higher values) with depth. Typically, however, clays hydrate and wash out in pressured zones, and borehole rugosity will affect the sonic values if it is severe, to the extent of causing cycle-skipping. Modern tools are self-compensating for hole washout, but the problem cannot totally be removed. A useful cross-check to see if the sonic values are representative is to correlate the values with those from a velocity analysis (refer to paragraph 4.2).

The theory behind quantitative geopressure evaluation using the sonic tool is fortunately independent of the amount of porosity. Note that the sonic log can be a definitive geopressure indicator: an increase in transit time, with depth, in constant clay lithology, is due to a change (increase) in porosity (pore pressure gradient). A quantitative interpretation is based on Gulf Coast methodology and may not be as accurate in other areas. However, it has been found to be a very useful tool.

As clays compact and lose porosity with depth, the measured sonic transit time will also decrease. Typical values for unconsolidated clays lie between 150-200 sec/ft. A plot of clay transit times on semilog grid produces a linear normal trend with depth. Hottman and Johnson (1965) correlated transit time deviations from the normal trend to adjacent reservoir pressures, but this method was specific only to some Gulf Coast reservoirs. The equivalent depth method can be used for most geopressure determinations from logs. Figure 4-25 illustrates the procedure.

The normal trend is extrapolated to the depth of interest. (Note: on placing the normal trend, remember that the minimum transit time is equal to the matrix transit time, i.e. zero porosity. Since rocks of zero porosity rarely exist, the slope of the normal gradient must intercept at values greater than the matrix transit time.)

The formation pressure is then determined:

$$P = OBG_a * D_a - D_n(OBG_n - N.FBG) \qquad (4\text{-}24)$$

where

$P$ = pore pressure (psi)

$OBG_a$ = overburden pressure gradient at $D_a$ (psi/ft)

$OBG_n$ = overburden pressure gradient at $D_n$ (psi/ft)

$D_a$ = depth of interest in geopressure (ft)

$D_n$ = normal equivalent depth (ft)

$N.FBG$ = normal formation balance gradient (psi/ft)

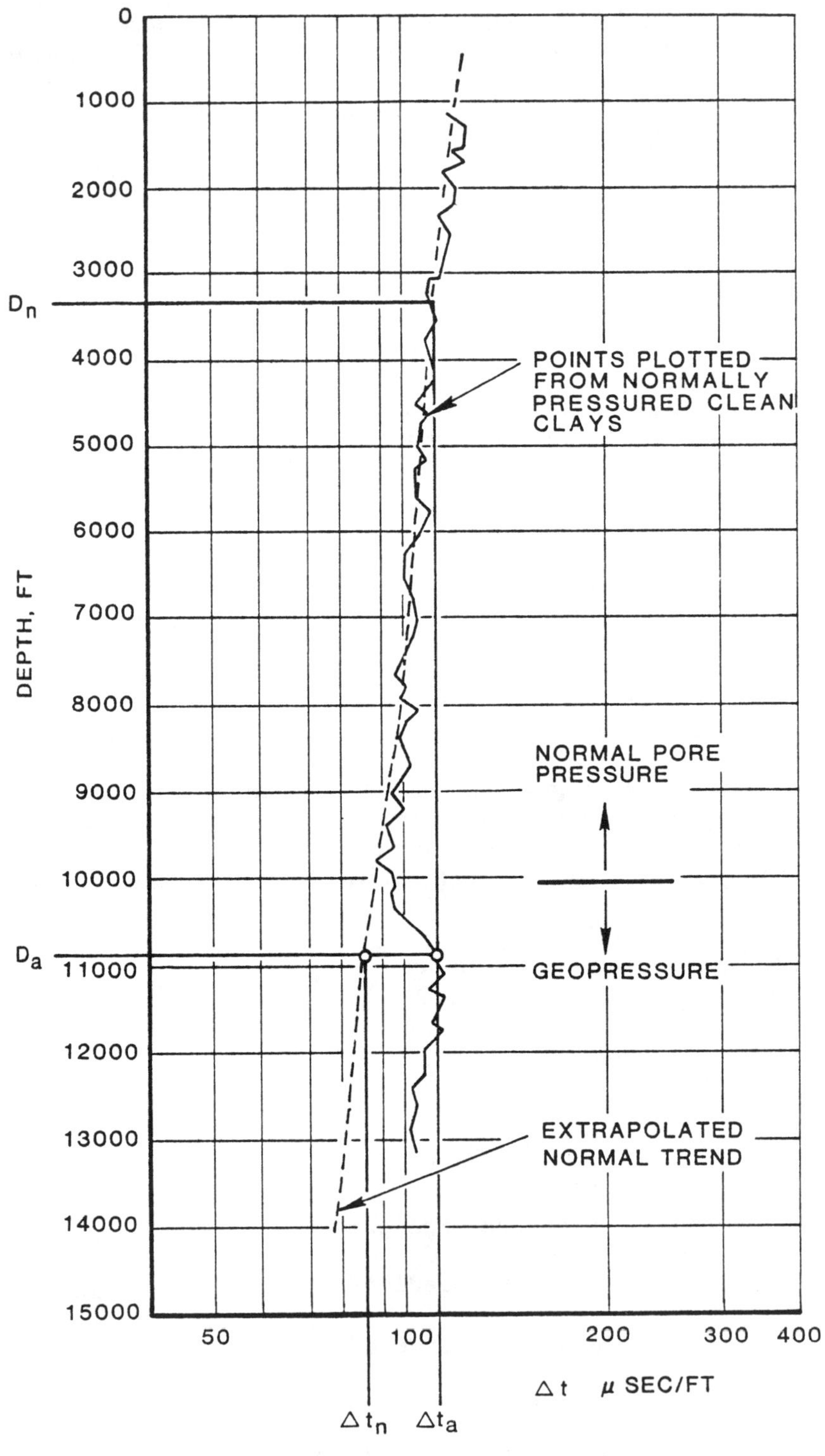


Figure 4-25. Geopressure evaluation using the equivalent depth method and sonic plot

For example (from Figure 4-25),

$$OBG_a = 0.920 \text{ psi/ft at } D_a$$

$$OBG_n = 0.830 \text{ psi/ft at } D_n$$

$$N.FBG = 0.465 \text{ psi/ft}$$

$$P = 0.92 * 10870 - 3300 \quad (0.83 - 0.465)$$

$$P = 8796 \text{ psi, or } 15.6 \text{ lb/gal at } D_a$$

Another application for the sonic log is its use for estimating overburden gradients and pressures through porosities and bulk densities. AGIP engineers Bellotti and Giacca (1978) published an empirical relationship that enabled bulk densities to be established directly from transit times. By calculating porosity from transit times a matrix and fluid are assumed, the densities corresponding to that particular matrix and fluid are known, and hence a bulk density may be estimated (see paragraph 3.4).

As porosities vary with lithological change and pore pressure, calculations should be made in each lithology with depth. Note that all rocks contribute to the overburden pressure: not just clays, hence readings should be taken from all rock types.

To obtain bulk densities, perform the following:

1. Identify lithological changes from the mud log (or other log) and correlate to sonic log.
2. Average transit times for each interval (eyeball is sufficient).
3. Identify matrix and obtain matrix transit time from Figure 4-24 for each interval.

For each of these intervals:

4. Calculate porosity using Equation (4-23) and using 189 μsec/ft for fluid transit time. Suggested matrix transit times for very shallow, wet clays; sub-compact clays and compact clays are 100, 70, and 65 μsec/ft, respectively. Local experience may dictate otherwise.
5. Correlate matrix used with matrix and fluid density from Figure 3-3.
6. Calculate bulk density using

$$\rho_b = \phi\rho_f + (1 - \phi)\rho_m \tag{3-8}$$

where

$\rho_b$ = bulk density (g/cc)

$\phi$ = porosity (fractional)

$\rho_m$ = matrix density (g/cc)

$\rho_f$ = fluid density (g/cc)

7. Overburden pressure gradients may then be calculated using Equations (3-10) and (3-11), or by using the EAP routine.

Calculated porosities will be abnormally high in clay intervals that have hydrated due to mud reaction. As the depth of investigation of the sonic tool is extremely small (less than 1 inch beyond the borehole face), hydrated clays will be measured, not the true formation. Care must be taken in these situations as a low calculated overburden gradient may result; however, no quantitative correction is available.

## 4.17 RESISTIVITY

The common rock-forming minerals conduct very little electricity. They effectively have zero conductivity. The resistivity of a rock is dependent upon the amount of water, its salinity, hydrocarbons, and the distribution of the fluids within the rock's porous network. Thus changes of porosity, water resistivity, hydrocarbon content, and porosity distribution within the same rock will cause changes in the resistivity measured by the various tools. Resistivity tools currently used are

- Normal and lateral types
- Micrologs
- Focused resistivity types
- Induction devices

and each has a particular use and application in log analysis. For geopressure evaluation purposes, the best logs to use are the induction and microlog types. The induction log is intended to read $C_t$, the true conductivity of the undisturbed formation. Values from this log are thus a function of porosity, porosity distribution and water salinity. The microlog measures two areas: (1) a microlateral which is affected by mud cake, and (2) a micronormal that measures the resistivity of the flushed zone. Since the resistivity of the mud filtrate is known and is thus the same for all flushed formations (temperature corrected), the resistivity of the flushed zone, $R_{xo}$, is thus a function of porosity and pore geometry only. However, the latter device is restricted in its use to formations of greater than 5% porosity and less than 0.5 inch of filter cake.

Whatever device is available, Gulf Coast experience has shown that a resistivity plot on semilog scale from clean clay beds produces an increasing trend with depth. The manner in which this trend increases is not predictable in wildcat areas, but in areas of intensive drilling, normal resistivity trends should be available. Typical normal trends are shown in Figure 4-26.

The porosity increase in geopressured clays is reflected by a decrease in resistivity, provided the resistivity of the pore water has not increased. The latter proviso is not predictable in wildcat areas, hence the resistivity log must be used with caution, both as a geopressure indicator and evaluation tool. In well known areas, however, the resistivity device has been found to be a reliable indicator and quantifier.

Calculation of pore pressures are facilitated by the use of Equation (4-24), the equivalent depth expression, where deviations from the normal resistivity trend are utilized instead of transit times.

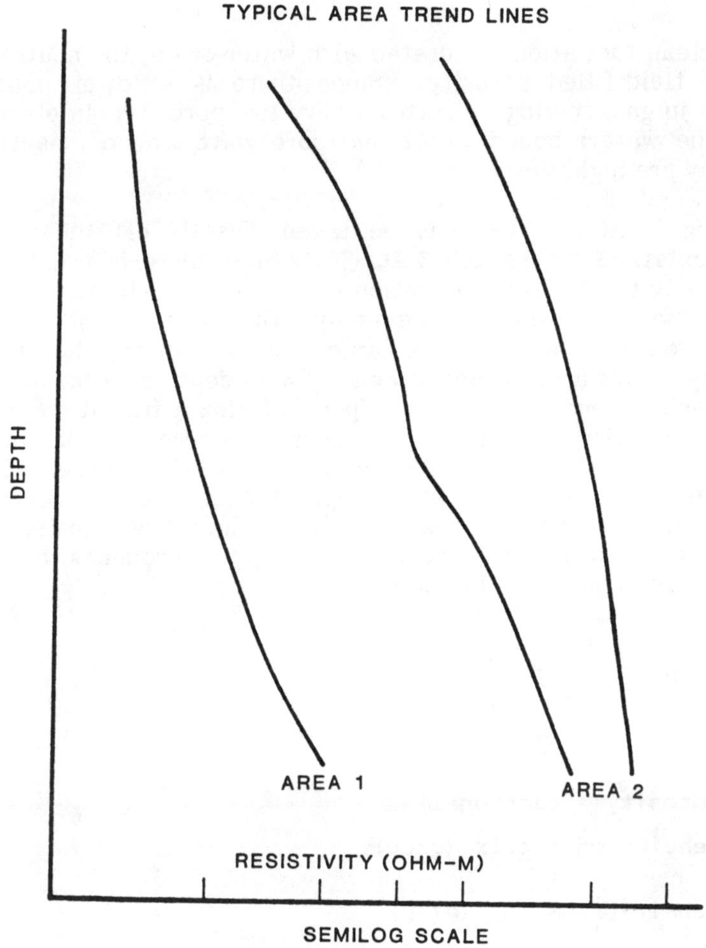


Figure 4-26. Formation resistivity typical area trend lines

## 4.18 DENSITY AND NEUTRON LOGS

The density log (FDC, Densilog etc.) measures formation density by bombarding the formation with gamma rays from a cesium 137 source, and detecting the energy and amount of radiated gamma rays from the formation. If it is assumed that the Mass Absorption Coefficient is constant for all rocks and fluids at a specific energy level, then the amount and relative energy of returning gamma rays is a measure of the density of the material.

The neutron log bombards the formation with highly energetic neutrons. The neutrons gradually lose energy as they migrate from the source, and at a very low energy level they are captured by nuclei of the formation. The detector on the tool senses gamma radiation of absorption or low energy neutrons. The greatest energy is lost when neutrons collide with a hydrogen nucleus: they have similar mass. Hence, the slowing of neutrons depends largely on the amount of hydrogen in the formation. In clean formations saturated with water or oil, the neutron log reflects the amount of fluid-filled porosity. Since there is a lower concentration of hydrogen atoms in gas, the log indicates a very low porosity. In clays, the neutron log reads all the water: bound water and pore water, hence, neutron porosities measured in clay are high.

If a density log is run, values may be taken directly off the log and used in overburden calculations (paragraph 3.4). Note that there is usually a correction scale in the density track. This correction is one that has already been applied, and it is plotted for the sake of completeness only. The correction should be seen to be greatest in washed-out hole, and the larger the correction, the less reliable the resultant density values are. A plot of density with depth on a linear grid displays a gradually increasing trend with depth. Upon entering a transition or geopressured zone in clay, the density curve may be seen to decrease. If the lithology is constant, this is a definitive indication of porosity increase. The depth of investigation of a density tool is about eight inches into the formation, hence, hydrated clays will affect the readings by causing low density values to be recorded. Calculation of porosity from the density tool produces the most accurate values overall. Use Equation (3-9) and Figure 3-3.

$$\emptyset = \frac{\rho_m - \rho_b}{\rho_m - \rho_f} \tag{3-9}$$

where

$\emptyset$ = porosity (fractional)

$\rho_m$ = density of matrix (g/cc)

$\rho_b$ = density from log (g/cc)

$\rho_f$ = fluid density (g/cc)

Pore pressures can be calculated in a geopressured zone using the density log with readings taken in clean clays. The equivalent depth method (Equation (4-24) and Figure 4-25) is used in the same manner as that described for the sonic log.

The neutron log is not used as a geopressure indicator or quantifier; however, it may show changes in clay porosity index that may be used to indicate a predominant geopressure mechanism. Montmorillonite clays will cause rapid neutron adsorption due to their very high bound water content; hence, the porosity index will be very large. Illitic clays have much less adsorbed water, hence, the porosity index should be correspondingly lower. Reference to Section 2 shows that clay composition changes through a geopressured section, depending on the predominant mechanism. Neutron response may indicate

- Compaction disequilibrium: clays within the geopressured zone are immature relative to shallower clays, hence, the neutron porosity index may increase markedly within the zone

- Montmorillonite dehydration: clays in geopressured zones have changed to illite, releasing water to the pores; much of this water must be released otherwise the pore pressure balances the overburden and any subsequent increase will promote the formation of horizontal fractures, allowing the pressure to dissipate. The neutron response would thus be constant through the zone, or a sharp decrease at the top of the geopressure if the excess water had been released

- Aquathermal: since this process involves compaction disequilibrium, the neutron response will increase within the geopressured zone

## 4.19 FACTORS AFFECTING EVALUATION

### 4.20 LITHOLOGY

The classic sand/shale sequences of typically marine sediments are perhaps the easiest to evaluate for geopressures. Lithological alternation is displayed in the Dxc plot, total gas plot, cavings occurrence, temperature plot and, of course, the wireline logs. Thick shale sequences allow normal trend development, permeable sandstones provide good pore pressure estimations by mud density/gas relationships, and geopressure trends are displayed in clay intervals.

Massively thick clays provide excellent opportunities for drilling exponent evaluation and cavings analysis.

Massively thick sands/arenaceous lithologies cannot be evaluated by textbook exponent methods; great thicknesses of turbidites, greywacke, tuffs and terrigenous clastics with few intercalated argillaceous horizons exclude the possibility of developing clay normal trends; however, these sediment types do exhibit normal trends in Dxc, density, and temperature plots. As most of the geopressure evaluation techniques are based on clay analyses, arenaceous lithologies severely restrict evaluation methods. Nevertheless, differential pressure is a major clue for evaluation, and mud density/gas relationships must be utilized to the utmost. With these tools, geopressure evaluation can be achieved with confidence, albeit at a degree of detail somewhat less than that possible in argillites. Where permeability is restricted in arenaceous sediments, the possibility of a geopressure occurring below becomes increased. The change in differential pressure upon entering a higher pore pressure zone will be monitored by the Dxc: the Dxc will decrease, as it would in geopressured clays. Evaluation, however, may not be of the same order as that in clays. The ratio method (Equation 4-11) will provide a very vague estimate, as empirical justification for pressure evaluation in sand types is not yet available. Thus, geopressure techniques may be used in thick sands, but the emphasis of evaluation must be shifted from Dxc analysis to mud density/gas/differential pressure methods, that is a qualitative evaluation of the magnitude of under- or over-balance.

Geopressure evaluation in carbonates can be the most difficult and frustrating task. Carbonate sediments can encompass the whole gamut of porosity range, permeability range, and pore geometries from huge caverns through open fractures to secondary solution types in microfossils. The characteristic variability of carbonates causes concomitant variability in geopressure plots. Argillaceous limestone and calcareous claystones can be evaluated (in the majority of cases) in the same manner as clays; i.e., all the evaluation techniques apply. Clastic limestones, without a high degree of cementation, may be evaluated as sand type sequences; i.e., increase emphasis on differential pressure evaluation techniques. Well cemented, massive types; i.e., micrite, secondary cemented fossiliferous limestone etc., can be extremely difficult to evaluate. These limestone types have highly variable permeabilities, and this is what makes things difficult. If a limestone is without permeability, a transition zone cannot exist. Totally impermeable types (porous, but the pores are not connected) may have extremely high pore pressures, probably caused by aquathermal mechanisms, but should not cause major drilling problems due to their impermeability. It is the permeability barrier below which is a highly pressured porous zone that provides the greatest potential danger.

As cemented limestones have a relatively high tensile strength, cavings do not appear until the degree of underbalance becomes high. Changes in differential pressure will affect the Dxc, but to an unpredictable extent. Bulk density measurements on cuttings should reflect the actual density, as hydration problems do not occur; hence, density measurements in limestone will indicate porosity changes; but due to the competent nature of the rock, a porosity increase does not necessarily indicate a concomitant pore pressure increase.

Probably the only techniques of anticipating the probable occurrence of a geopressure within highly competent, massive limestones, are the various temperature plots. Again, however, the assumption is that the geopressured interval will be porous and water-filled, so that it may act as an insulator to heat. If a temperature gradient reversal does occur with depth, it may be assumed that a zone of considerable porosity (fluid-filled) is being approached, but this could either be a fractured, vugular, dolomitized or granular interval of high or low pore pressure. In any event, drilling should proceed with great caution until the character of the anomaly is determined. (See paragraphs 5.3 to 5.5 for lost-circulation problems.)

Apart from the above discussion, there can be no hard-and-fast rules laid down for geopressure evaluation in carbonates. Areal experience should be developed in areas of intensive drilling in carbonates; but at all times, drilling rank wildcat wells in carbonate lithologies calls for the most diligent observations and interpretations on the part of the geologist.

## 4.21 CONTROLLED DRILLING

Controlled drilling offshore in top-hole sediments is commonplace and desirable. It does, however, cause problems with exponent analysis. Because the bit is not wholly drilling, and the jetting action of the bit is the major drilling mechanism, drilling exponents cannot be used in their accepted role. Consider, however, that as Dxc is a differential pressure indicator, no matter what the drilling mechanism is, rate of penetration rises as differential pressure decreases. Then as the penetration rate is controlled, and rotary speed is kept constant, Dxc becomes largely a function of bit weight which is allowed to vary. Bit weight thus changes with formation type and character. In soft, unconsolidated clays, jetting will proceed with vigor and will be considerably aided with increasing pore pressure. Rate of penetration being controlled allows bit weight to reduce to negligible quantities, causing deflections to the left on the Dxc plot.

Thus in soft top-hole sediments, geopressure indications may well be exhibited by the Dxc plot, and this is of particular importance in attempting to ascertain the presence of shallow, pressured, gas reservoirs (see paragraphs 4.4 and 4.5). Pore pressure quantification cannot be performed from the Dxc plot in these situations, because

- Establishment of a normal trend in top-hole is difficult
- The bit is not truly drilling, hence, the Dxc values are not indicative of actual "drilling" values
- Deviation of the Dxc points to the left may indicate increasing pore pressure, but the ratio method cannot be applied in unconsolidated sediments

Mud density/gas relationships should provide a reasonably accurate estimate of pore pressure magnitude and changes.

Whether the overall drill rate is controlled by slow penetration or by circulating between singles, the shallow gas pocket may make its presence known without warning. The client and drilling personnel must be made aware of the limitations of geopressure evaluation in shallow sediments, particularly the quantitative aspect. An unconsolidated sand containing pressured biogenic gas will not be heralded by a transition zone: the surprise element thus becomes magnified to startling proportions, so for safe tophole drilling, drilling crew diligence must be tuned accordingly.

## 4.22 HYDRAULICS

Rates of penetration, and hence drilling exponents, are a function of the various hydraulic forces at the bit. Current exponents do not take into account the effects of changing hydraulic parameters: they assume hydraulics are at all times 100 percent efficient and optimized. Pump efficiency, surface pressure losses and the various down-hole frictional losses may be calculated, but there is really no very accurate method of measurement at present, in order that calculations may be checked.

Inefficient drilling hydraulics suppress rates of penetration and cause inflated exponent values. Overly energetic hydraulics promote washouts, pump failure, increased bit wear and hole problems. The optimum conditions are between 60 and 70 percent of total hydraulic horsepower for maximum hydraulic horsepower at the bit; but for maximum jet impact force, hydraulic horsepower at the bit should be approximately 50 percent of the total.

The different hydraulics involved in turbo and diamond drilling contribute to shifts in the various trends, but how much of the shift is due to the change in drilling mechanism alone cannot be determined.

## 4.23 BIT SELECTION AND BIT WEAR

If a bit selection is made such that it cannot drill the formation efficiently, exponent trend response will be considerably masked. The most typical error is opting for an insert bit in moderately hard formation, only to find that the bit produces very sluggish rates of penetration. Cases are known where a geopressure transition zone was drilled with an inefficient bit; the result was a normal or slightly shallower Dxc trend, completely masking the increasing pore pressure. If this is permitted to continue, loss of the hole could occur due to sudden sloughing, or a kick could be taken.

The decision to change from a milled-tooth to an insert bit is a difficult one, particularly in wildcat areas. When the change is made, geopressure indicators other than drilling exponents should be monitored with increased concentration, as the situation could be such that the bit masks a transition zone.

New bit selection is partly dependent on the amount of wear that the previous bit sustained. Unfortunately, the accepted "eyeball" technology so favored by drilling crews can be so affected by extraneous phenonena that the result recorded on the drilling report may bear little resemblance to the bit in question. By using a simple quantitative method, bit grading becomes meaningful and may be gainfully employed in calculating Nx parameters. The method is not rigorous, but provides consistency and reasonable accuracy in the time-frame available.

The suggested method is as follows (see Figure 4-27).

1. Before the new bit is run, measure the height of one tooth on each row (use number 2 or 3 cone)

2. Count the number of teeth on each row of that cone, and multiply the tooth height pertaining to that row by the number of teeth in that row.

3. Multiply this result by the number of cones.

Note that most bits have positional tooth and row variations, so the result will not be the actual total tooth height.

When the old bit becomes available, perform the same measurements on the same cone, and calculate the quantitative tooth wear as shown in Figure 4-27.

## 4.24 MUD TYPE

The mud type must be compatible with the formation. Specific muds may be developed for specific wells, so the formation reaction, reservoir interaction, temperature effects and rig problems can be minimized. Often it is not possible for one mud to achieve all these goals, and it is common for muds to be changed for a particular well section. Again, the problems are accentuated in wildcat areas. Complete information can be obtained from the various manuals that the mud companies provide, so it is sufficient here to simply outline possible occurrences that could hinder geopressure interpretation.

NEW BIT

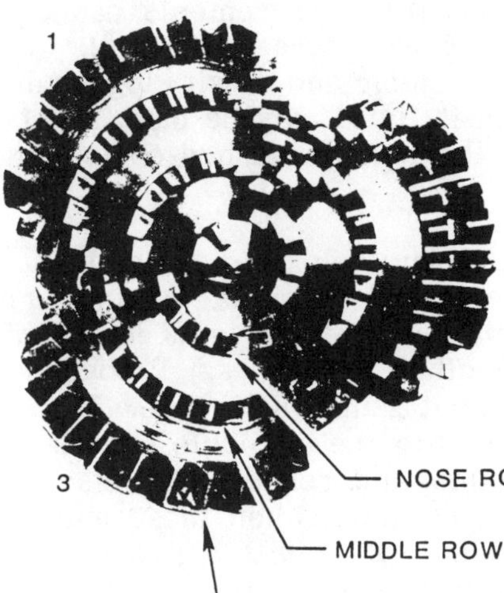


USING NUMBER 3 CONE

HEEL ROW : 24 TEETH
TOOTH HEIGHT : 1"
TOTAL HEIGHT : 24"

MIDDLE ROW : 18 TEETH
TOOTH HEIGHT :0.8"
TOTAL HEIGHT : 14.4 "

NOSE ROW : 10 TEETH
TOOTH HEIGHT : 0.8"
TOTAL HEIGHT : 8"

TOTAL TOOTH HEIGHT FOR CONE : 46.4"
TOTAL TOOTH HEIGHT FOR BIT :
3 x 46.4" = 139.2"

WORN BIT

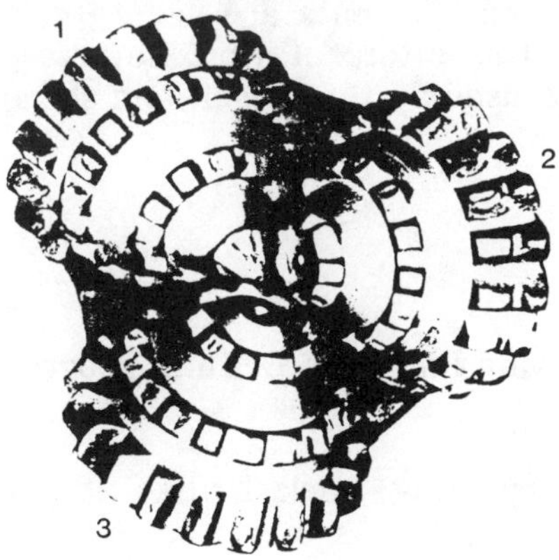


USING NUMBER 3 CONE

HEEL ROW :22 TEETH (2 MISSING)
TOOTH HEIGHT : .7"
TOTAL HEIGHT : 15.4"

MIDDLE ROW : 18 TEETH
TOOTH HEIGHT : .5"
TOTAL HEIGHT : 9"

NOSE ROW : 9 TEETH (1 MISSING)
TOOTH HEIGHT : .5"
TOTAL HEIGHT : 4.5"

TOTAL TOOTH HEIGHT FOR CONE : 28.9"
TOTAL TOOTH HEIGHT FOR BIT :
3 x 28.9" = 86.7"

BIT WEAR = (139.2 − 86.7) ÷ 139.2 x 100 = 37.7%

TOOTH GRADE = 8 x 37.7 ÷ 100 = T3

Figure 4-27. Estimation of Bit Tooth Wear

Water-based muds, fresh or salt, may react with hydrateable clays. If reaction does occur, clay cuttings swell, lose their morphology and even dissolve in the mud. The result is rapidly increasing mud density and solids content, increasing viscosity and a distinct absence of clay cuttings. Bulk density measurements cannot be performed on this "gumbo," but shale factors can (see paragraph 4.12). Clay cavings from a transition zone may not be apparent. Sonic and density log readings in clay zones could also be anomalous: sonic transit times may be high, and bulk densities may be extremely low, particularly if the hole is washed out which is usual in these situations.

Inhibitive muds are types that combat clay swelling and help to diminish hole and mud problems. Calcium, gypsum, spersene, saturated salt (NaCl), and ligno-sulphonate types all control clay hydration to various extents, and choice of a particular mud type is made depending on their other properties. Probably the most effective clay inhibitor water-based mud is a potassium (KCl) type. This mud serves to provide potassium cations for adsorption onto the available lattice sites of montmorillonite, collapses the expanded lattice and renders the clay non-reactive. Good clay cuttings and cavings can be obtained when these muds are used. Shale factor values from clays that have been drilled with a KCl mud type should be considerably less than the actual original exchange capacity; if the KCl system is kept efficient, change in clay mineralogy may be completely masked and shale factor trends rendered meaningless. However, adsorption of potassium by montmorillonite never seems total, and some degree of hydration occurs.

Diesel-based muds are by far the best drilling fluid for aiding geopressure evaluation. All cuttings and cavings are preserved in their original form. No hydration occurs, shale factors and bulk density measurements are accurate, and sonic and density log curves are representative. Gas interpretation, however, is made more difficult due to the background level caused by the diesel, and slugs of fresh diesel may cause further problems.

## 4.25 REFERENCES

Bellotti, P., and D. Giacca, 1978, Pressure Evaluation Improves Drilling Programs, O&GJ, Sept. 11

Bingham, M. G., 1965, A New Approach to Interpreting Rock Drillability, The Petroleum Publishing Co.

Dowdle, W. L., and W. M. Cobb, 1975, Static Formation Temperature from Well Logs, J. P. T., Nov.

EXLOG, 1985, Mud Logging: Principles and Interpretations, IHRDC Press, Boston.

Fowler, P.T., 1980, Telling Live Basins from Dead Ones by Temperature, World Oil, May.

Goldsmith, R.G., 1972, Why Gas-cut Mud is Not Always a Serious Problem, World Oil, v. 175, n. 5.

Hottman, C.E., and R.K. Johnson, 1965, Estimation of Formation Pressures from Log-Derived Shale Properties, J.P.T., Jun.

Jorden, J. R., and O. J. Shirley, 1966, Application of Drilling Performance Data to Overpressure Detection, J. P. T., Nov.

Lewis, C. R., and S. C. Rose, 1970, A Theory Relating High Temperatures and Overpressures, J. P. T., Jan.

Nwachukwu, S. O., 1976, Approximate Geothermal Gradients in Niger Delta Sedimentary Basin, AAPG Bull., v. 60, n. 7.

Pennebaker, E.S., 1968, An Engineering Interpretation of Seismic Data, SPE 2165, presented at the 43rd Ann Fall Meeting of SPE-AIME, Houston.

Rehm, B., and R. McClendon, 1971, Measurement of Formation Pressure from Drilling Data, SPE 3601, SPE Reprint Series No. 6a, 1973 revision.

Reynolds, E.B., 1970, Predicting Overpressured Zones with Seismic Data, World Oil, v. 171, Oct.

# 5

# LOST CIRCULATION, HYDRAULIC FRACTURING, AND KICKS

## 5.1 INTRODUCTION

Generally, the speed and efficiency with which a well is drilled are dependent upon the formation balance gradient/mud density relationship. The safety factor also depends primarily on this relationship. With the costs of wells continually escalating (particularly offshore), drilling time and material costs are minimized through engineering practices which attempt to produce maximum penetration rates as cheaply as possible. Instances have arisen where safety margins were negated when preference was given to cost/time activities rather than to established safety levels. Because safety should always come first, knowledge of kick tolerance is of paramount importance.

When drilling rank wildcat wells, predrill information is nonexistent or at best scarce and open to question. Accurate measurements and pressure interpretations on these wells are necessities. Drilling development wells or delineation wells within a known province may remove the surprise element to some degree, but this should never be taken for granted.

The economical aspect of drilling a well is of some concern to all types of Exlog logging units: GEMDAS, for drilling efficiency and optimization; pressure evaluation units, for more involvement with formation interactions; and the standard logging units, for lithological and gas interpretations. However, the safety aspect of drilling a well is of prime importance for all types of units. Recognizing underbalanced conditions, reporting unexplained pit level changes, and careful setting and rapid response to PVT alarms may be an important contribution to rig safety.

Generally, the safest and most economical well is that in which the estimated kick tolerance is never exceeded. In order to calculate the kick tolerance, it is necessary to know the following parameters: potential or actual lost circulation zones, hydraulic fracture pressures, and pore pressures.

## 5.2 LOST CIRCULATION

### 5.3 CAUSES

Lost circulation occurs when whole mud is taken by the formation. The rate at which mud is lost is dependent upon the type of formation and the mud density, and knowing this gives some idea of the severity of the situation. There are six major causes for lost circulation:

- The bit has penetrated a cavernous, vuggy formation
- The bit has penetrated open fractures or faults that have access to a lower pressure potential
- The circulating pressure of the mud has exceeded the fracture pressure of a formation
- Very poor hole-cleaning, resulting in a pack-off in the annulus. Mud pressure rises until fracturing occurs below the pack-off
- A zone of subnormal pore pressure has been penetrated so that either the formation has been fractured (paragraph 5.7) or the significant overbalance has brought about mud loss through massive filtration in the permeable formation
- The formation fractured while tripping in rapidly, or casing was run at excessive rates

Other mud losses, which are less drastic but would lead to hole problems if they are left unchecked, do not involve wholesale losses but rather are a result of excessive filtration. This is due to:

- High overbalance
- High fluid loss/mud chemistry
- Weak filter cake
- Highly permeable formation

The result is a continual slight mud loss while drilling, and excessive hole fill-up on trips. Extensive permeability reduction in potential reservoirs, termed "skin damage," is a result of mud plugging and filtrate interaction with sensitive clays in the pore spaces. Unchecked, this can render false reservoir parameters during testing; hence, all effort must be made to counteract this process.

## 5.4 EFFECTS

Depending on the rate and mechanism of mud loss, the usual effect is a rapid loss of varying degree: returns may cease completely, or may be reduced. If a vuggy formation is penetrated, and communication exists, the effective volume of the macroporosity may be so great that no volume of mud may fill it. Mud losses will continue until preventive measures are taken.

A faulted or jointed formation with considerable fracture permeability causes varying rates of mud loss, depending on the permeability and the fluid pressure potential between the fractured formation and the borehole. However, the reverse may be true: fluid pressures in the fractures may be higher than the pressure in the borehole, and the well may kick. Usually, if an extensively fractured formation is encountered, mud loss is extremely rapid and will not cease until preventive measures are performed.

If the mud density is high in relation to the pore pressure and overburden pressure, the formation could be fractured. Mud loss is rapid, but circulation may easily be restored by dropping the mud density until the pressure reduction allows the fractures to close. This was a problem in the early days of wildcatting, and was overcome by a reduction in mud density. Unintentional hydraulic fracturing resulting in major fluid loss occurs less frequently in modern drilling. However, experimental work with the borehole televiewer (a downhole sidewall sonar device) indicates that most boreholes have some degree of minor hydraulic fracturing, probably caused by pressure surges when running pipe.

If a subnormal pore pressure zone is encountered there is a possibility that hydraulic fracturing will occur, due to the fact that a pore pressure reduction will produce a fall in fracture pressures (paragraph 5.24).

Continual hole fill-up and increased hook loads while tripping in are indications that the formation has been fractured at some point below the bit or last casing shoe (if running casing). This is caused by the combination of mud density and surge pressures. Usually the fractures will close when the trip is completed or when surge pressures are minimized. Losing circulation while running casing is particularly hazardous because a poor cement job may result, allowing communication behind the casing. Many blowouts have occurred after cementing casing, and losing circulation while running or cementing the casing is often a contributing factor.

## 5.5 SOLUTIONS

Rapid and continual mud loss while circulating can thus be caused by two different mechanisms: (1) fracturing, and (2) loss through interconnected vugs or preexisting open fractures. The first mechanism may be arrested by slowing the pump rate (thus lowering the ECD) or by changing the mud properties if the fracture pressure has been slightly exceeded; or alternatively, a reduction in mud density might be necessary. The latter mechanism requires the addition of lost-circulation material (LCM) to attempt to bridge the vugs. If this fails, then a cement squeeze operation may be necessary.

Lost returns into a highly fractured formation may be minimized by injecting pellets or sand of decreasing size, so that fractures become bridged and packed. If this succeeds, then increasingly-fine injected material will improve the ability of the packed pellets to reduce flow. Ultimately a mud filter cake may form, allowing normal drilling to resume without further losses. If packing and bridging of the fractures are unsuccessful, then the interval must be cemented off and redrilled.

Penetration into a zone of subnormal pressure may cause significant problems other than formation fracturing. These zones are permeable, so pipe sticking is a real danger. It will be necessary to reduce the mud density as much as possible, taking into account the open hole above the low pressure zone. Depending on their permeabilities, the higher formations could kick or could slough severely due to the decrease in mud support. If the magnitude of the pressure reduction in the permeable formation is too high, such that further drilling could cause increased borehole instability, it would be necessary to seal the zone with cement. In such circumstances this operation is difficult because the cement may flash-set, producing no improvement. An impermeable seal, however, must be made before drilling can safely be resumed. If all else fails casing must be run, and this may necessitate several cementing operations.

Mud loss to the formation due to vugs or open fractures should not be confused with fracture pressures. If the formation is such that a high pressure potential exists between the borehole and the fluids in the crack or vug porosity, then mud loss will occur (providing there is ample permeability) until the pressure potential is equalized. Normally, if the vugs and fractures are interconnected, the volume necessary for pressure equalization is far in excess of the available mud supply. In this case, returns will not be gained until the thief zone is sealed off or the mud density is reduced so that it equals the fluid pressure in the cracks. Also, if a normal-pressured vuggy or fractured formation has exceedingly high permeability, then due to the enormous volume available in the formation, a mud at very slightly higher pressure will preferentially flow into this formation. No fracturing is involved; the formation acts as a sponge.

The formation will actually fracture within fairly well-defined narrow limits if all the necessary parameters are accurate. Thus formations that are thief zones (e.g. a 10 lb/gal mud is continually lost and loss continues even when the mud density is reduced to 8.6 lb/gal) due to enormous fracture or vug porosity will have normal fracture pressures depending on the actual pore pressure, rock type and overburden pressures. Mud loss occurs when the mud pressure exceeds the fluid pressure in the fractures or vugs. The actual pore pressure in the rock itself will be very similar to the fluid pressure in the crack unless the fracturing (i.e. fault brecciations) occurred recently and the permeability of the whole rock is such that pore pressure/pore fracture pressure equilibrium has yet to occur. Thus in thief zones, a fracture pressure for that formation will be meaningless unless the flow zones are sealed.

If lost circulation does occur, <u>every attempt must be made to keep the hole full</u> by continuous addition of mud or water. Allowing the hydrostatic pressure to fall below the pore pressure in a permeable formation may result in a kick or an underground blowout which will be exceedingly difficult to control.

In summary, lost circulation zones have enormous permeability and porosity, and mud loss will continue until the mud pressure in the borehole equals the fluid pressure in the cracks or vugs. If the borehole pressure falls below the fluid pressure, the flow will reverse. When the thief zone is sealed by plugging and filter cake, mud loss will cease and mud densities can then be raised to a value below the estimated fracture pressure, without further loss.

## 5.6 HYDRAULIC FRACTURING

## 5.7 PAST AND CURRENT TECHNOLOGY

The hydraulic fracturing technique of well stimulation has been used since the 1940s. Prior to the mid-1950s, both this process and the similar, costly and time-consuming occurrence of lost circulation while drilling with high mud densities were thought to occur due to the formation of horizontal or bedding plane fractures. Lifting the overburden in this manner was the explanation put forward, totally disregarding the fact that most of the pressures in the borehole at the time of fracturing were considerably less than the total weight of the overburden. Theoretical studies and accurate pressure measurements made during squeeze cementing operations raised questions as to the validity of this argument of lost circulation, due to the formation of horizontal fractures; pressures required in boreholes were mostly less than those of the overburden, so the orientation of the fractures were inferred to be vertical. In 1949, Clark (of Stanolind Oil & Gas Co.) showed how flow through hydraulic fractures could be greatly increased by pumping in sand with the fluid. The sand prevented the fractures from closing, thus providing a conduit from the reservoir to the well. At this time there existed amongst engineers a stubborn standing that all fractures were bedding-plane fractures. The Shell Oil Company, in 1955, employed M. K. Hubbert to provide a critical review of the situation, and the result was the classic paper, "Mechanics of Hydraulic Fracturing," published in 1957.

Through the use of accepted engineering theory, Hubbert and Willis showed that a subsurface stress regime is such that, when normal faults occur ($60^{o}$ to the horizontal), the minimum horizontal compressive stress is of the order of one-third to one-half of the maximum vertical compressive stress.

(In a subsurface environment, there exists a system of stresses. At any point in that environment, the stresses acting on that point can be resolved into three mutually perpendicular stresses: a maximum, intermediate, and minimum stress, $\sigma_1$, $\sigma_2$ and $\sigma_3$, respectively. Geologists use the notation that compressive stresses are positive, engineers use the convention that tensile stresses are positive; hence in the latter case, $\sigma_3$ is the maximum compressive stress. Stress is pressure, or force per unit area, and always acts normal to a selected plane. If the simplest subsurface environment is taken (i.e. horizontal beds, horizontal topography, elastic rocks, and horizontal constraint), the maximum compressive stress ($\sigma_1$) is vertical and equal to the pressure of the overlying rocks; and since the rocks have been assumed to be isotropic, the horizontal stresses will be equal and will act in all directions in a horizontal plane, and are caused by a function of Poisson's ratio of the rock type and $\sigma_1$. If a superposed horizontal stress is imposed on the system (i.e. a tectonic stress), then the horizontal stresses will become unequal and directional such that $\sigma_2$ is parallel to the tectonic stress and $\sigma_3$ is normal to $\sigma_2$ in the horizontal plane. This is illustrated in Figure 5-1. When pressure is applied in a borehole, it creates tensile stresses around the walls. If the tensile stress exceeds the horizontal compressive stress in the surrounding rocks and also overcomes their

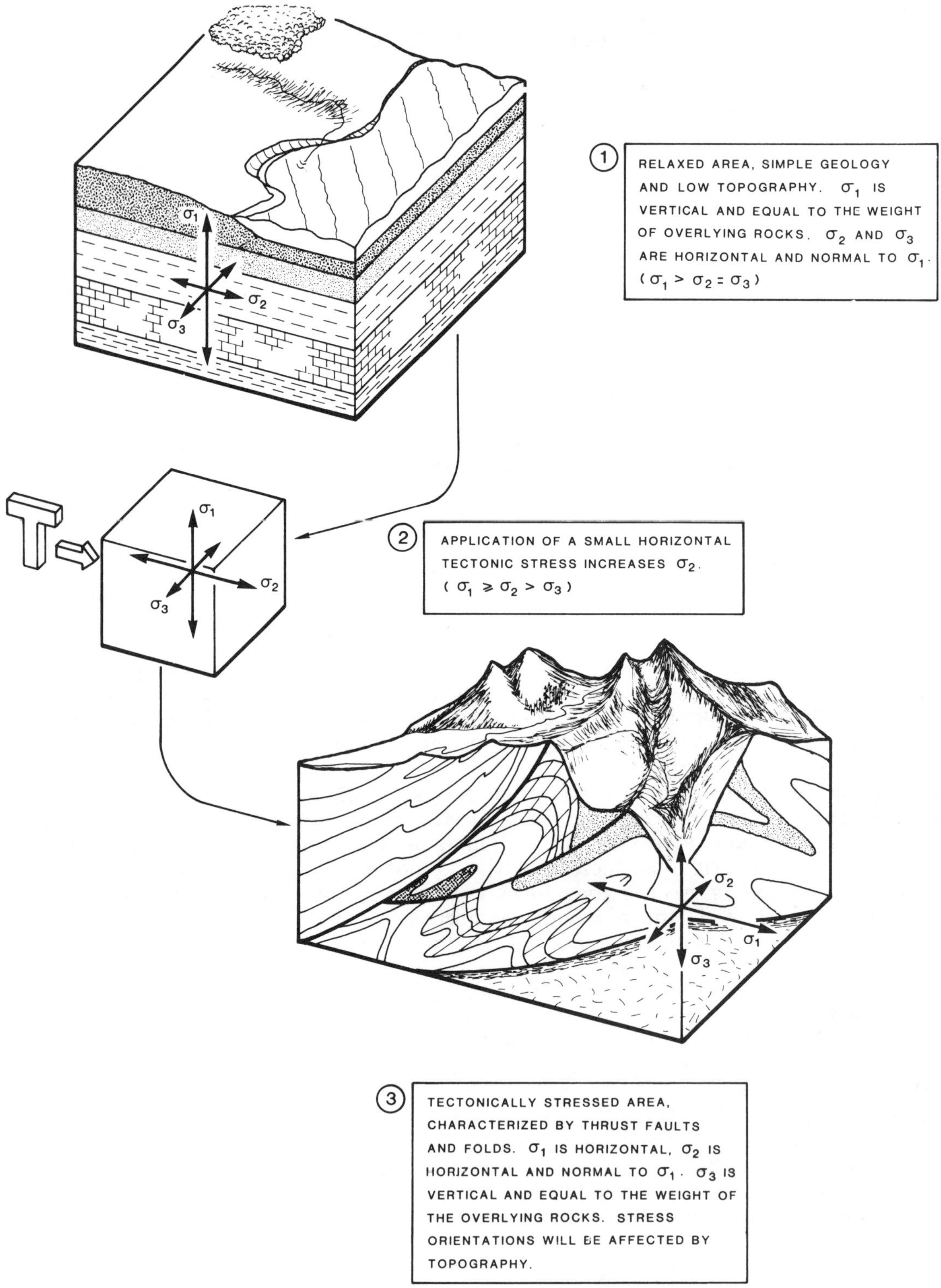


Figure 5-1. Stress regimes in relaxed and tectonic areas

tensile strength, a tensile fracture will form along the path of minimum resistance, i.e. normal to $\sigma_3$ and parallel to $\sigma_2$ and $\sigma_1$. If $\sigma_1$ is vertical (if the basin is relaxed), the tensile fractures will be vertical and oriented parallel to $\sigma_2$ if $\sigma_2$ is greater than $\sigma_3$. If a superposed tectonic stress is imposed such that it is of greater magnitude than the pressure of the overburden, then $\sigma_1$ is horizontal and parallel to the tectonic stress and $\sigma_3$ is vertical. To cause fracture in this case, the pressure in the borehole must be slightly in excess of the total pressure of the overburden, and the fracture will be horizontal.)

Hubbert and Willis overcame the problem of attempting the predictions of the tensile strengths of rocks in situ by observing that many closed cracks, joints and partings intersect any section of borehole. The effective tensile strength of the rocks over the interval are thus close to zero. When an interval is hydraulically fractured, the pressure in the borehole must balance the minimum stress holding the preexisting cracks closed, and must provide a further small amount of energy to extend the cracks. If a crack exists in a compressive stress field and a pressure is applied within the crack such that it balances the compressive stress acting normal to the sides of the crack, a further slight increase in crack pressure will produce an exceedingly high tensile stress at the tip. This tensile stress easily overcomes the tensile strength of the rock, and the crack rapidly propagates.

Utilizing these assumptions, Hubbert and Willis showed that fracture will occur when

$$F = \frac{(S-P)}{3} + P \qquad (5\text{-}1)$$

where

F = pressure in the borehole at point of fracture (psi)

S = total pressure of the overburden (psi)

P = pore pressure (psi)

Thus the minimum injection pressure required per unit depth in an area of incipient normal faulting is

$$\frac{F}{D} \approx \frac{\frac{S}{D} + \frac{2P}{D}}{3} \qquad (5\text{-}2)$$

where

D = depth (ft)

This expression provided an estimate for the minimum fracture pressures that occur in a relaxed basin that is on the point of normal faulting. Hubbert and Willis concluded that fracture pressures are affected (1) by the magnitude of the preexisting regional stresses, (2) by the hole geometry (including any preexisting fissures), and (3) by the penetrating quality of the fluid. Further, to simplify the calculation, they assumed that if the value of S/D is approximately equal to 1 psi/ft and under normal hydrostatic conditions (P/D) of 0.46 psi/ft, the minimum fracture pressures to expect in the Gulf Coast would be 0.64 psi/ft.

Hubbert and Willis' paper thus provided the theoretical and technical basis for predicting minimum fracture pressures, as well as a means to predict fracture pressures in tectonic environments and abnormal pressure zones, if the relevant parameters could be measured. However, this was not sufficient for the industry since wells drilled in areas of active normal faulting are very few and far between. The need to predict fracture pressures at any point in a borehole became necessary to plan casing programs — especially in areas where, due to high pore pressures and/or tectonic stresses, abnormal hole conditions were usual.

In 1967, Matthews and Kelly published a study in which fracture pressures could be predicted in some Gulf Coast sand reservoirs by the use of empirical data. Since this area was undergoing extensive exploration, their data allowed safer and more economical well completions. Unfortunately, Matthews and Kelly did not further the progress made by Hubbert and Willis. They chose the minimum fracture pressure as being equal to the pore pressure, and the maximum fracture pressure equal to the pressure of the overburden. A fracture pressure that was observed to be greater than the pore pressure was thought to be due to the force necessary to overcome the "matrix load" or the "cohesive nature of the matrix." Then, "using the assumption that the cohesive property of the matrix may be related to the matrix stress and hence will vary only with the degree of compaction, the following relationship has been developed for calculating the fracture gradient of sedimentary formations." (This is illustrated in Figure 5-2.)

$$\frac{F}{D} = \frac{P}{D} + ki\frac{\sigma}{D} \tag{5-3}$$

where

F = fracture pressure (psi)

P = formation-fluid pressure at depth of interest (psi)

D = depth at point of interest (ft)

$\sigma$ = matrix stress at the point of interest (psi)

ki = matrix stress coefficient for the depth at which the value of would be the normal matrix stress

In developing their method, Matthews and Kelly assumed that the average normal hydrostatic gradient is 0.465 psi/ft and that the average overburden gradient is 1.0 psi/ft. In abnormal pressure zones, the increase in pore pressure (P) produces a corresponding decrease in the matrix stress ($\sigma$), since

$$\sigma = S - P \tag{5-4}$$

where

S = overburden pressure

The value for ki is taken from the depth at which $\sigma$ is normal (Figure 5-3). These empirical values and relationships are limited solely to the area of study.

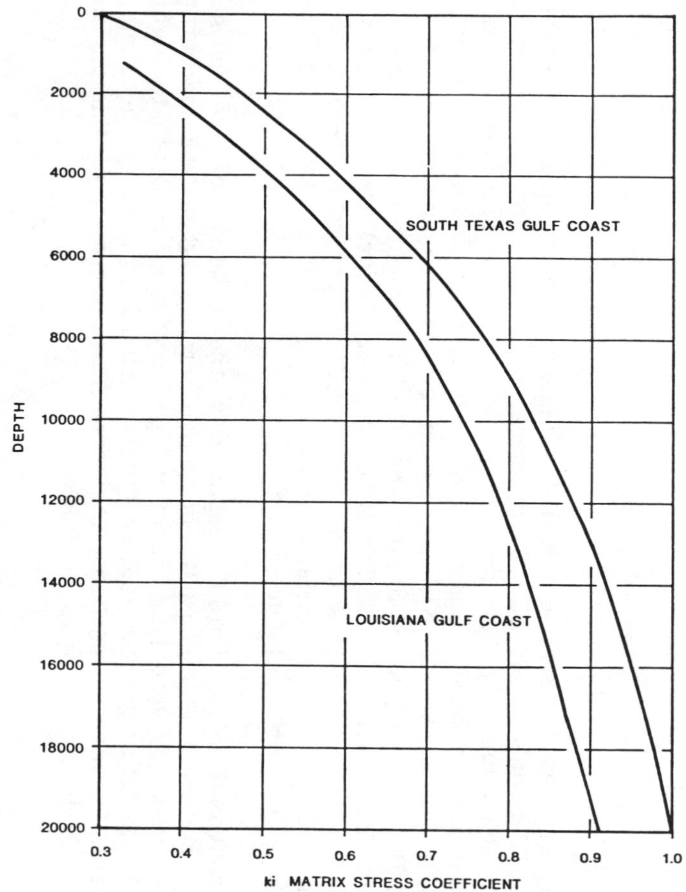


Figure 5-2. Matrix stress coefficient (ki) with depth for Gulf Coast Sands (Matthews and Kelly, 1967)

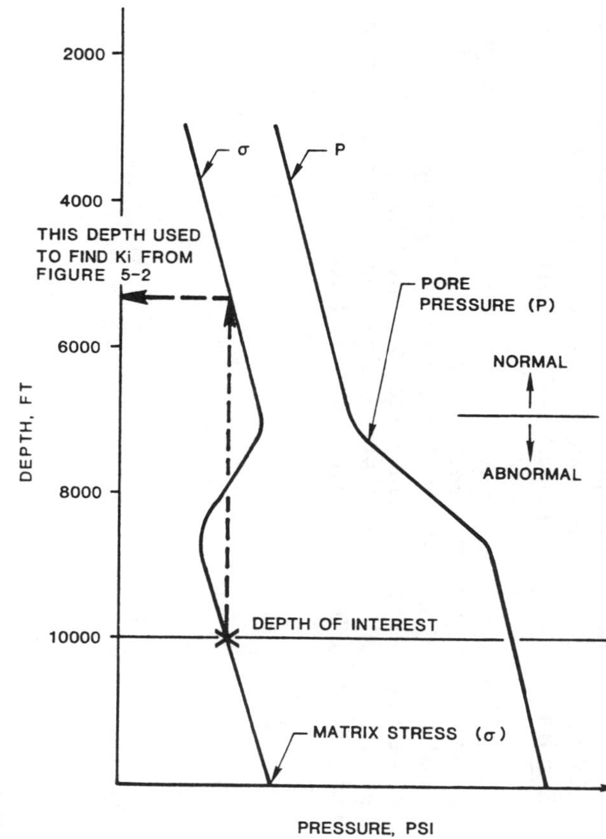


Figure 5-3. ki is obtained from the depth at which $\sigma$ is normal

Further empirical studies were published by Costley (1967) who utilized similar ideas. In 1969, Eaton published a more adaptable method that took into account a variable overburden gradient. Eaton introduced Poisson's ratio as a variable that controlled fracture pressure gradient. Poisson's ratio ( $\mu$ ) is formally defined as

> The ratio of the lateral unit strain to the longitudinal strain in a body that has been stressed longitudinally within its elastic limit. It is an elastic constant.

It is thus a property of the rock itself. Eaton surmounted the problem of predicting or measuring Poisson's ratio of every in-situ rock in a borehole by resorting to an empirical relationship. Further, Eaton's "Poisson's ratio" is not a function of the rock itself but of the regional stress field — the horizontal-to-vertical stress ratio. Thus, as Hubbert and Willis assumed that the minimum horizontal stress would be $\sigma_1/3$, this corresponds to a "Poisson's ratio" of 0.25 through the relationship

$$\frac{F}{D} = \left(\frac{S}{D} - \frac{P}{D}\right)\left(\frac{\mu}{1 - \mu}\right) + \frac{P}{D} \tag{5-5}$$

when

$$\frac{S}{D} = 1.0 \text{ psi/ft}$$

and

$$= 0.25$$

Then

$$\frac{F}{D} = \frac{\left(1.0 + \frac{2P}{D}\right)}{3} \tag{5-6}$$

which is the same as Hubbert and Willis' minimum fracture gradient relation. However, the "Poisson's ratio" of 0.25 predicts values that are usually too low compared with the value from field data; also, the assumption that S/D = 1.0 psi/ft leads to error (except in West Texas wells where fracture gradients are a minimum as predicted by Hubbert and Willis). Eaton presented empirical curves for "Poisson's ratio" versus depth calculated from Gulf Coast data. The curves with depth approach the upper limit of 0.5; that is, a longitudinal strain produces an equal lateral strain: this occurs in materials with a zero shear modulus (e.g. liquids) and in incompressible materials. These curves, which are thus independent of rock type, are illustrated in Figure 5-4.

The major contribution of Eaton's paper was the concept of the variable overburden. The assumption of 1 psi/ft for an overburden gradient was inaccurate; gradients were found to vary from about 0.6 psi/ft at shallow depth to slightly greater than 1.0 psi/ft at greater depths. Since overburden pressures play a major role in fracture gradient estimations, the increase in accuracy of this variable allowed better fracture gradient estimations.

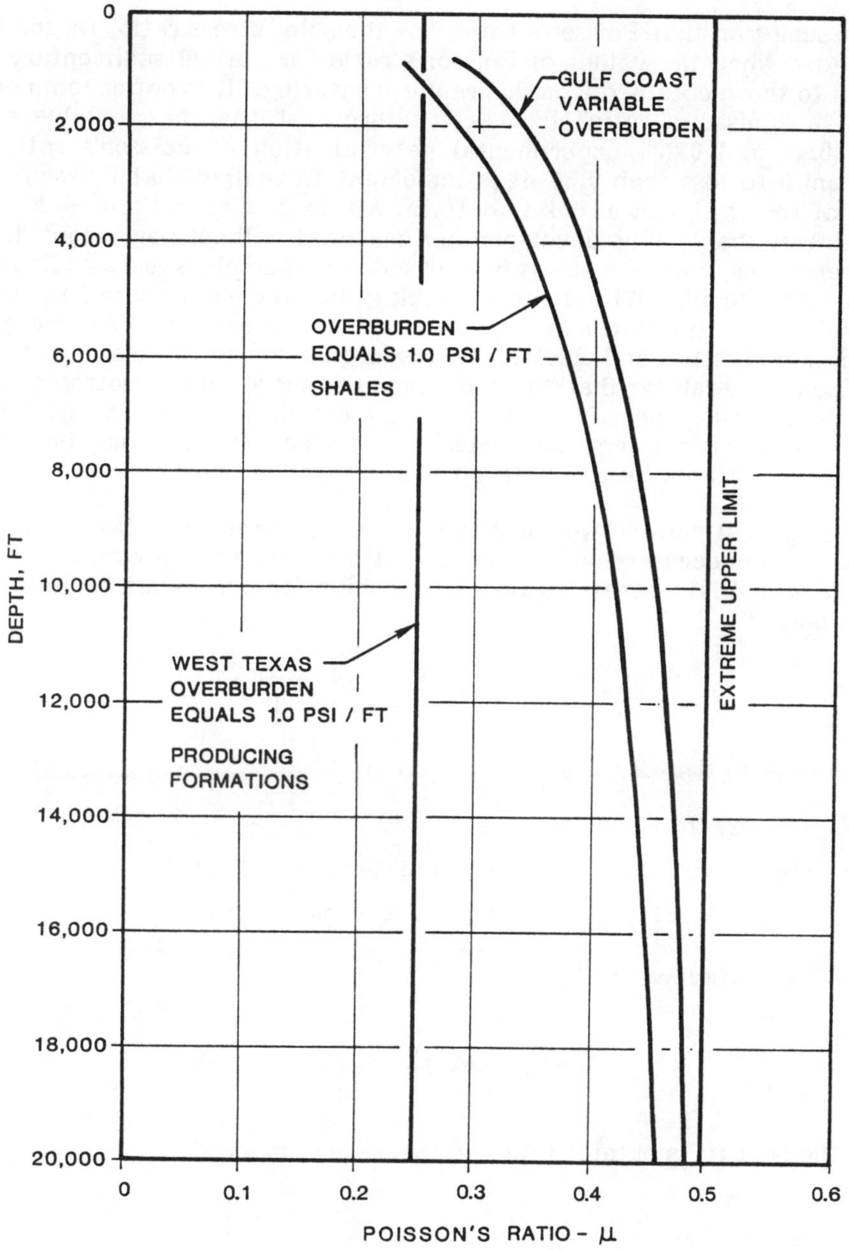


Figure 5-4. Empirical "Poisson's ratio" curves with depth for Gulf Coast sands (Eaton, 1969)

Eaton's technique can be applied in other areas if the "Poisson's ratio" curve is known. Thus it is limited to areas of concentrated exploration in tectonically relaxed regions and cannot be used reliably on wildcat wells.

Eaton's assumption that Poisson's ratio was the sole "stress ratio" factor appears to be unfounded when the values of Poisson's ratio for normal sedimentary rocks are compared to those obtained from hydraulic fracturing. It is not uncommon to back-calculate a "Poisson's ratio" from a fracture test that has a value somewhere between 0.45 and 0.8. Experimental determination of Poisson's ratio produces values from 0 to less than 0.5. It is important to realize that Poisson's ratio is a measure of the ability of a rock to deform, within its elastic limit — defined as the greatest stress than can be developed in a material without permanent deformation (strain) remaining when the stress is released. Surface clays are usually so wet that they behave as liquids. With depth the rock grains themselves are responsible for a unique Poisson's ratio, but, as compaction increases, the rocks become more dense and solid, more brittle, and elastic. This is largely due to the closure of cracks and creep of the minerals so that the rock becomes increasingly isotropic with depth. Since elastic rocks transmit seismic energy efficiently, and "plastic" rocks may transmit compressional acoustic waves but not shear waves, it may be realized that "plastic" rocks will not be encountered within drillable depths.

Another empirical method was published by Anderson et al in 1973. Their aim was to derive all the necessary parameters to estimate fracture pressures from electric logs. Utilizing Biot's stress/strain relationships for porous media, they developed the relationship

$$\frac{F}{D} = \left(\frac{2\mu}{1-\mu}\right) * \frac{S}{D} + \left(\frac{1-3\mu}{1-\mu}\right) * \frac{\alpha P}{D} \qquad (5\text{-}7)$$

where

$\alpha$ = $1 - Cr/Cb$

Cr = compressibility of the solid matrix material

Cb = compressibility of the porous rock skeleton

can be approximated by

$$\alpha = 1 - (1 - \varnothing_D)^n \qquad (5\text{-}8)$$

If n = 1, the best fit is obtained for the theoretical models.

Hence,

$$\alpha \approx \varnothing_D \qquad (5\text{-}9)$$

Hence, $\alpha$ is also dependent upon porosity, but is an immeasurable quantity in a drilling environment. Terzaghi experimentally found that if $\alpha = 1$, then the relationship becomes

$$\frac{F}{D} = \left(\frac{2\mu}{1-\mu}\right) * \frac{S}{D} + \left(\frac{1-3\mu}{1-\mu}\right) * \frac{P}{D} \qquad (5\text{-}10)$$

which is thus independent of porosity.

But the problem still remains for obtaining $\mu$ for in-situ rocks. Theoretically, $\mu$ can be obtained from the sonic shear and compressional velocities (Vs and Vc) in the formation by the relationship

$$\mu = \frac{1 - 2\left(\frac{Vs}{Vc}\right)^2}{2\left(1 - \frac{Vs}{Vc}\right)^2} \tag{5-11}$$

The recognition of shear wave arrivals in most sedimentary sections is usually impossible. In order to obtain Poisson's ratio for Gulf Coast sands, Anderson et al made the broad assumption that "Poisson's ratio is a function of the shaliness of the sand since the shale would act essentially as a plastic bonding agent." The estimation of the shale content of the sand from sonic and density logs was accomplished by the use of a shale index:

$$I_{sh} = \frac{\phi_S - \phi_D}{\phi_S} \tag{5-12}$$

where

$I_{sh}$ = Shaliness index

$\phi_S$ = porosity from sonic log

$\phi_D$ = porosity from density log

For shaliness index from 0 to 40 percent, Poisson's ratio was found to vary from 0.27 to 0.33 for the Gulf Coast sands. This linear relationship can be solved for $\mu$ :

$$\mu = A * I_{sh} + B \tag{5-13}$$

where

$I_{sh}$ = shaliness index

A = the slope of the line

B = the intercept on the y axis

This relationship has been developed only for the data collected (i.e., from Gulf Coast sands), and obviously other relationships occur in other sands with different clay, clay structure, sand/clay relationship, and sand types.

Christman, in 1973, accented the problem of assuming a 1-psi/ft overburden gradient when drilling offshore. On offshore rigs, a high flowline elevation above sea level and drilling in deep water were shown to cause important modifications to calculated overburden and other pressure gradients. Refer to Section 3, "Overburden Pressure", paragraph 3.4, for a complete discussion.

Recently, Bradley (1979 a,b) published a complicated theoretical concept that could provide limits for borehole stability when a significant angle exists between the borehole and the regional stresses. Limits are set for failure in compression (sloughing) and failure in tension (fracture). Due to the very large number of variables involved, a computer is used to calculate and plot all the possible states of stress for all hole angles and directions. The result is an area, or "stress cloud." Any change in variable will produce a change in shape of the stress cloud and a movement of the cloud across the mean shear stress/mean normal stress plane. A failure envelope experimentally obtained from rock failure at different confining pressures defines the limit of stability within the "stress cloud." An application of this model may be of use on development platforms where deviated wells are drilled.

## 5.8 LIMITATIONS AND ADVANTAGES OF ACCEPTED MODELS

### 5.9 Hubbert and Willis's Minimum Fracture Gradient

The fact that this theoretical model does not use empirical constants or relationships must be a point in its favor. Unfortunately, however, it appears that the industry has chosen to misinterpret the object of this work — that is, to provide a means by which minimum fracture gradients may be obtained. Also, the theory may be applied in any location, provided that all the provisions are met (i.e., an area characterized by normal faulting, simple topography, and horizontal beds). The main disadvantage of this model is that it is imprecise. When hole conditions are such that very accurate fracture gradients are necessary, a minimum value is not sufficient.

### 5.10 Matthews and Kelly's Method

Application of this model is limited to the Gulf Coast area since it was developed on wells in the Gulf Coast (specifically, in producing sands). The empirical value, ki, may be back-calculated from a succession of fracture tests in an area, and then curves must be plotted against depth. Matthews and Kelly's curve cannot be used because they assumed a 1-psi/ft overburden gradient and because it relates to the Gulf Coast reservoir sands. This method can be used only within a single field in which sufficient fracture data is available to plot a ki curve which will be unique to that field.

### 5.11 Eaton's Method

Eaton attempted to define the problem of determining the actual subsurface stress regime by use of "Poisson's ratio." Basically, the reasoning is precisely the same as Hubbert and Willis' except that Eaton endeavors to account for a higher-than-minimum horizontal stress. He found empirically that, with a variable overburden gradient, their "stress ratio" ( $\sigma_3 / \sigma_1$) or ki varied nonlinearly with depth. As with Matthews and Kelly's method, ki curves have to be back-calculated from a multitude of data within a single field before predictions can be made.

## 5.12 Anderson et al

Again, Poisson's ratio is a necessary variable; however, in this model the ratio is a function of the rock itself, and not a "stress ratio" independent of rock type. Because of the difficulty of recognizing shear arrivals on sonic logs, is empirically related to the percent of clay in the reservoir sands of the Gulf Coast. Also, the rock compressibility parameter, , is defined by a relationship which in turn is empirically related to these sands; specifically,

$$\alpha = 1 - (1-\emptyset_D)^n \qquad (5\text{-}14)$$

where

n = 1, and gives a best fit to the data

$\emptyset_D$ = porosity from the density log

If n = 1, the relation is approximated to $\alpha = \emptyset_D$ which can be applied only to those particular sands. In combination with the questionable relationship between $\mu$ and shaliness, again, this method is limited to the area in which it was developed. Use in other areas will necessitate different $\mu$ and shaliness relationships to be developed, and possibly the determination for $\alpha$ will have to be reevaluated.

It must be noted that, specifically, the last two methods were developed for sandstones. Limestones, shales and other typical sedimentary rocks could produce spurious results simply because their properties were not considered.

## 5.13 ESTIMATION OF FRACTURE PRESSURES

With drilling now extending to deep waters and high latitudes, the costs of these wells are becoming exceedingly high. Deep wildcatting in areas of poor geological control can be extremely hazardous and costly for lack of adequate pore pressure and fracture pressure information. If abnormally high pore pressures are encountered, a further casing string may be necessary; and if the pressure zone is shallow in relation to the target, completion of the well could be jeopardized.

Of prime importance in these wells is an accurate assessment of kick tolerance. For this to be achieved, knowledge of the fracture pressures at any depth in the open hole is necessary. The prediction of fracture pressures in Gulf Coast fields and other areas that have been extensively drilled is accomplished by the use of empirical formulae. These can only be applied with confidence in other areas of similar geological and tectonic regime when sufficient drilling has allowed the calculations of the necessary empirical constants. However, the absence of any method by which fracture pressures may be predicted outside these areas has necessitated the use of these empirical formulae, with the general result that actual fracture pressures can be very different from calculated pressures. This is mainly due to the application of the empirically derived constants, usually representing the "stress ratio," which are unrelated to the wildcat area. Accurate information on the in-situ principal stresses is vital for the solution of the fracture pressure problem. None of the empirical formulae can accurately predict stresses in localized regions.

A hypothesis is proposed that has the capacity to resolve and extrapolate the local principal stresses, subsequent to the first fracture test in compact formation. Compact is defined here as the point at which the sediment can transmit an applied stress through the grain contacts. Along with other pertinent data usually calculated on rank wildcats, i.e. overburden gradients and pore pressures, fracture pressures can then be obtained for any point within the drilled hole. Kick tolerance calculations then become more realistic when they are based on fracture pressure calculations for that specific well, so that in the event that abnormal hole conditions are encountered, the chances of completing the well are greater than if reliance is placed on formulae containing unrelated empirical constants.

In order to hydraulically fracture the formation, it is necessary to overcome the minimum compressive stress. General formulae describe the minimum horizontal compressive effective stress as a function of the effective overburden pressure, which is empirically derived:

$$F = \sigma_3' + P \tag{5-15}$$

where

$F$ = fracture pressure

$P$ = pore pressure

$\sigma_3'$ = minimum compressive effective stress

and

$$\sigma_3' = K\,(S-P) \tag{5-16}$$

where

$K$ = empirical "stress ratio" constant

$S$ = overburden pressure

Overburden pressure is obtained by integrating bulk density with respect to depth:

$$S = \int_0^z \left(g * \rho\right) dz \qquad \text{(Jaeger and Cook, 1976)} \tag{5-17}$$

where

$g$ = acceleration due to gravity

$\rho$ = density

$z$ = depth

The in-situ stress regime can be calculated from

$$\sigma'_3 = \sigma_t + \sigma'_1 \left(\frac{\mu}{1 - \mu}\right) \tag{5-18}$$

where

$\sigma_t$ = superposed horizontal tectonic stress

$\sigma'_1$ = maximum compressive effective stress

$\mu$ = Poisson's ratio

and

$$\sigma'_1 = S - P \tag{5-19}$$

$$\sigma_t / \sigma'_1 = \beta \tag{5-20}$$

## 5.14 SUBSURFACE STRESS STATES

### 5.15 Effective Stresses

The concept of effective stresses was first introduced by Terzaghi in 1923 and has subsequently been used extensively in mechanical applications. Basically, a hydrostatic stress (P) within a pore fluid has no influence on deformation, which is controlled by the effective stresses. This hydrostatic stress is a "neutral" stress, one that acts in all directions and in the same amount. This stress is regarded to exist in both the solid and the liquid, so the effective stresses arise exclusively from the solid skeleton. Major studies on rock deformation (Handin et al, 1963) have shown that fracture is controlled by the effective stresses, provided the rocks have a connected pore system:

$$\sigma'_1 = \sigma_1 - P, \quad \sigma'_2 = \sigma_2 - P, \quad \sigma'_3 = \sigma_3 - P \tag{5-21}$$

where

$\sigma_1\ \sigma_2\ \sigma_3$ = principal maximum, intermediate and minimum compressive stresses

P = pore pressure

$\sigma'_1\ \sigma'_2\ \sigma'_3$ = principal compressive effective stresses

To apply this concept to a subsurface environment it must be assumed that the permeability is sufficient to allow movement of fluid and that the pore fluid is inert, so that the effects are purely mechanical.

To illustrate the effect of pore pressure on the vertical stress, let the total overburden pressure at 10,000 ft be 9500 psi, and the pore pressure at that depth be 4671 psi. The effective vertical stress is then 9500 - 4671 = 4829 psi. If the pore pressure at 10,000 ft was 8304 psi, then the effective vertical stress would be only 1196 psi.

## 5.16 Theoretical Subsurface Stress States

There are two major schools of thought regarding the state of stress within the earth's crust:

1. That the stress state is hydrostatic -- the three principal stresses are equal.

2. The horizontal principal stresses are a function of the effective vertical stress and Poisson's ratio.

The first hypothesis is generally termed Heim's rule and was later described by Anderson (1942) as the "standard state." It was stated in the form that stresses in rock tend to become equal because of the ability of the rocks to creep, such that any stress difference will eventually become alleviated. This hypothesis might be best illustrated by visualizing a scale model of the earth (Hubbert, 1945). Although the earth as a whole has the strength of cold steel, if it is modeled as a 4-ft-diameter sphere, it would have the strength of pancake batter and a viscosity about twice that of honey, and would weigh 6.6 tons.

The second hypothesis describes the state of stress in an elastic, flat-lying stratum of semi-infinite extent that is laterally constrained. If the weight of the overlying strata is the only source of stress, and the elongation in the horizontal directions are zero, then the relation

$$\sigma_H = \sigma_1'\left(\frac{\mu}{1-\mu}\right) \tag{5-22}$$

is derived, where $\sigma_H$ and $\sigma_1'$ represent the horizontal and vertical effective stress components, respectively, and $\mu$ is Poisson's ratio. If, for example, Poisson's ratio for a particular rock type is 0.25, then the horizontal stresses would be one-third that of the vertical stress, provided the theoretical conditions were satisfied. In contrast, Heim's rule states that the horizontal stresses should be equal to the vertical stress.

Common to both theoretical discussions are the assumptions that one principal total stress is vertical and equal to the weight per unit area of the overlying rocks, and that the horizontal normal total stress is the same in any direction in the horizontal principal plane.

That the crustal stress state is largely not hydrostatic is illustrated by the number of structures and deformation processes that necessitate unequal stress states for their formation and maintenance. Jeffreys (1952) suggested that significant stress differences occur within the upper 50 km of the earth's crust due to the existence of mountains and deep oceans. The occurrence of large-scale structures such as grabens, shear zones, dike swarms, nappes, folds, thrust and transcurrent faults suggest that not only did large stress differences occur in the past, but that stresses are still in a state of flux, as suggested by the occurrence of earthquakes. Some external stress, or tectonic stress, is necessary to produce these types of structures. Even in seismically inactive areas it is possible to infer a particular orientation of a tectonic stress, and it is reasonable to assume that even in the absence of tectonic structures and seismicity, a region may be subject to some tectonic stress (Jaeger and Cook, 1976).

Hafner (1951) showed that in order to obtain a hydrostatic type stress system (or "standard state") within a flat-lying stratum of infinite horizontal extent in which lateral extension is prevented, the stress system is composed of two parts:

1. The effect of gravity, described by the second hypothesis above

2. A superposed horizontal stress which is constant in any horizontal plane but increasing uniformly with depth

Moreover, for faulting and folding to occur, the superposed horizontal stress must occur in a particular orientation within the horizontal plane. If it exists, and as such would be a tectonic stress, it would also increase uniformly with depth, assuming that the strata were isotropic and elastic.

The horizontal stress can thus be a minimum when there is no tectonic stress, such that

$$\sigma'_3 = \sigma'_1\left(\frac{\mu}{1-\mu}\right) \tag{5-23}$$

where $\sigma'_3$ is the minimum principal horizontal effective stress, $\sigma'_1$ is the maximum principal effective stress which is equal to the effective pressure of the overlying rocks, and $\mu$ is Poisson's ratio for the particular rock type. The largest magnitude that the horizontal effective stresses can reach is approximately three times the vertical effective stress, at which point failure occurs in the form of reverse faulting (Hubbert, 1951).

The superposed horizontal tectonic stress, $\sigma_t$, can thus vary between the limits:

$$0 \leq \sigma_t < 3\,\sigma'_1 - \sigma'_1\left(\frac{\mu}{1-\mu}\right) \tag{5-24}$$

Since $\sigma'_1$ is calculated by subtracting the pore pressure from the total weight of the overlying strata, it is known for any point in the drilled hole. The superposed horizontal stress, if present, will increase uniformly with depth, or with $\sigma'_1$. Hence it may be assumed that the $\sigma_t/\sigma'_1$ ratio remains constant.

Ideally, Poisson's ratio for the rock type that is being drilled should be known at that moment in time, but this is not possible. However, Poisson's ratio has been experimentally measured for many rock types and is shown to be unique for a particular lithology. Poisson's ratio cannot be measured for each and every rock type, but if it is possible to divide lithological types into a grouping that can be described by a Poisson's ratio, then there exists a means by which experimental results may be applied to the same lithological types in situ.

To be able to describe the minimum horizontal stress, it is necessary to measure the magnitude of the superposed tectonic stress $\sigma_t$. This can be achieved by a fracture test. Hence, after $\sigma_t$ has been determined, the total horizontal minimum stress state can be extrapolated to any point in the drilled hole.

## 5.17 THE ZERO TENSILE STRENGTH CONCEPT

The accurate estimation of actual tensile strengths of subsurface sediments is probably impossible. Fortunately, this problem disappears if the common assumption, that any interval of sediment is intersected by joints and partings, is employed. Across these natural discontinuities the tensile strength is effectively zero. However, the occurrence of open joints or fissures is generally quite rare and may be restricted to a particular zone or lithology. Cracks in competent sediments form during compaction and diagenetic processes as a result of very localized stress differences. Microcracks are also formed due to the drilling process and the resultant stress-release at the borehole walls. Cracks that are held closed by the in-situ compressive stresses require a pressure within the borehole equal to the compressive stress, so that the pressure holding the crack closed is reduced to zero. A further slight increase in pressure in the borehole should allow entrance of fluid into the crack so that pressure is transmitted to the sides. This pressure will extend the crack indefinitely, provided it can be transmitted to the leading edge.

This phenomenon can be illustrated by considering a perfectly smooth, cylindrical borehole within an elastic medium, in which a crack extends to the wall of the hole. Upon an application of stress within the borehole that is slightly greater than the stress acting normal to the crack, a tensile stress is developed at the tip of the crack that approaches an infinite magnitude, as illustrated in Figure 5-5 (Hubbert and Willis, 1957).

The minimum pressure (F) necessary within the borehole to hold open and extend an existing fracture is therefore very slightly in excess of the regional horizontal stress normal to the plane of the fracture:

$$F = \sigma_t + \sigma_1'\left(\frac{\mu}{1-\mu}\right) + P \tag{5-25}$$

where

P = pore pressure

The plane along which a fracture will start to form will be that plane across which the compressive stress is a minimum, and thus will first be reduced to zero with increasing pressure in the borehole. In the case where the horizontal compressive stress is less than the vertical compressive stress, this plane will be vertical; if the horizontal stresses are greater than the vertical stress, the plane would be horizontal.

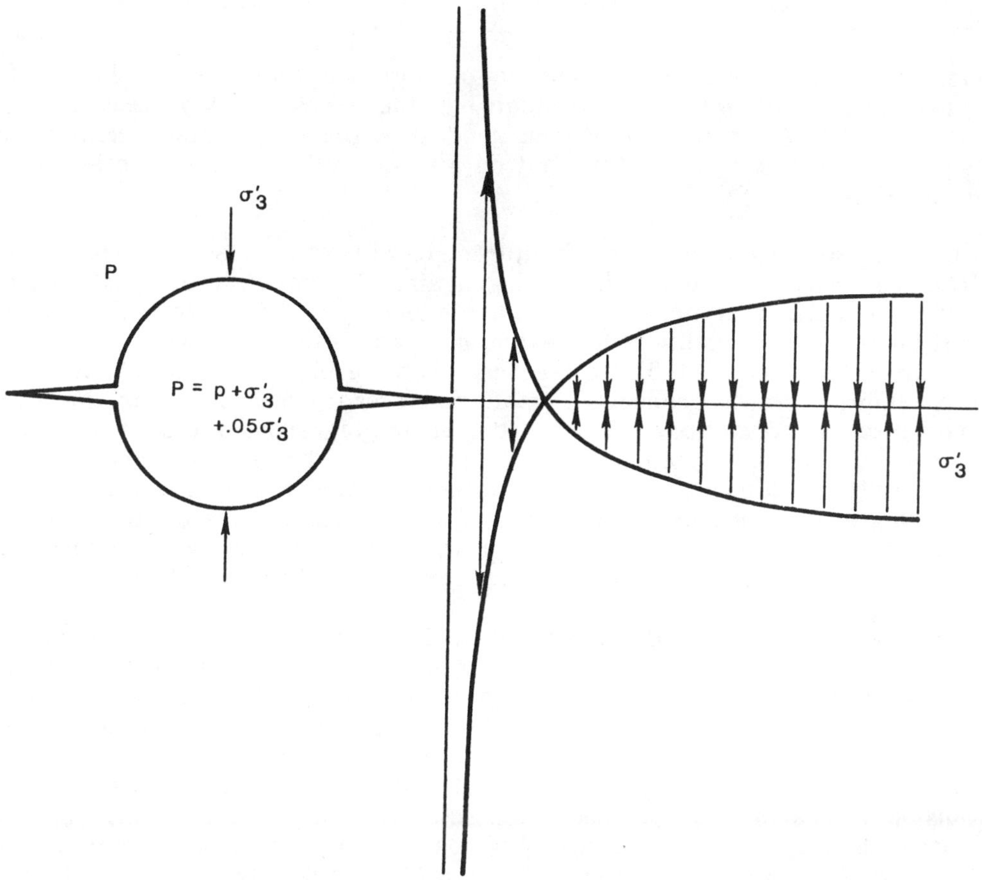


Figure 5-5. Extremely high tensile stress is produced at the tip of a crack when the pressure within the borehole is approximately 5% greater than the minimum horizontal stress (Hubbert and Willis, 1957)

## 5.18 THE FRACTURE TEST

Fracture tests are normally conducted after setting casing. The result of this test, when converted to an equivalent mud density, is taken to be the maximum mud density that the next hole section can withstand without losing circulation. Examination of the principles involved suggests that this assumption is valid only in a certain set of circumstances. If the last casing shoe was cemented in an abnormally high pore pressure zone and the pore pressure gradient then decreases significantly with depth, the fracture pressure gradient will decrease also. Limestone has a high Poisson's ratio, which will result in a higher fracture pressure than if the casing shoe was set in a rock with a lower Poisson's ratio. Drilling out of a limestone into a sand at the same or lower pore pressure gradient will result in a lower fracture pressure gradient.

Generally, the point in any section of borehole that has the lowest fracture pressure gradient will be that which has the lowest pore pressure gradient and lowest Poisson's ratio. Maximum mud densities for further drilling are thus dependent on these parameters, not on a unique value that was determined at the casing shoe.

Once the formation has been fractured, it will be necessary to apply that same fracture pressure to cause fracturing again. On any fracture test, the point at which the horizontal stresses become balanced by the pressure within the borehole will be the same, whether the test is a repeat or not. However, if a permeable formation is being tested, the fracture pressure plot will probably not be linear: the volume increase produces a smaller pressure increase, due to the invasion of fluid into the formation. This has the effect of raising the pore pressure of the formation immediately adjacent to the borehole. The increase in pore pressure has the result of reducing the stress concentration at the borehole wall, in turn resulting in a lower pressure necessary for fracturing. Once the fracture is started and is extending into the undisturbed stress field, the pressure required for this extension is the same as if no invasion occurred (Hubbert and Willis, 1957).

Fracture tests conducted offshore at shallow depths in unconsolidated clays can produce apparently abnormally high fracture pressures. Wet clays may behave as liquids, so that Poisson's ratio would be approaching 0.5. Also, as the pore water and adsorbed water may surround each clay platelet, the platelets will not themselves be in contact with one another, but will be supported by the water. These clay types have negligible shear strength. The effective pore pressure would thus be approaching the pressure exerted by the weight of the overlying sediments; when combined with a very high Poisson's ratio, it will be seen that calculated fracture pressures may exceed the overburden pressure by a significant amount. In these instances a horizontal fracture will form, lifting the overburden, so that the fracture pressure will approximately be equal to the overburden pressure.

At some depth, the weight of the overburden will squeeze out sufficient pore water so that the clay platelets become in contact with one another. When this occurs, the sediment can support a superposed horizontal stress. Poisson's ratio for the clay at this stage will be very similar to that of a more compact clay. Fracture tests in a clay which is at this stage of dewatering can be used for the calculation of the horizontal stresses.

Unconsolidated sands at shallow depths having a very good permeability may cause lost-circulation problems. Although the sand may be unconsolidated, the individual grains will be in contact so that a superposed stress can be supported independent of the pore pressure. Poisson's ratio will be normal, depending on the sand type. Consider that if an unconsolidated sand is drilled at 2000 ft, the overburden pressure is 1453 psi, and the pore pressure is normal at 892 psi. For a fossiliferous sandstone, Poisson's ratio is 0.01. Assuming that the horizontal stress ratio is normal, i.e. $\sigma_t/\sigma_1'$ is 0.2, then the calculated fracture pressure for these parameters is

$$F = \left[(1453-892)0.2 + (1453-892)\frac{0.01}{0.99}\right] + 892$$

$$F = 1010 \text{ psi, or } 9.7 \text{ lb/gal at } 2000 \text{ ft}$$

It can be seen that in shallow, unconsolidated sediments with high water content, normally encountered offshore, fracture pressures can vary from overburden magnitudes in wet clays to only a little more than the pore pressure in unconsolidated sands.

A typical fracture-test plot is shown in Figure 5-6. The linear portion of the curve, AB, indicates elastic properties: pressure increase (stress) is directly proportional to volume pumped (strain). At point B, the pressure within the borehole is equal to the pore pressure plus the total minimum horizontal effective stress. All cracks, joints and partings within the section of borehole that is being tested, that lie on a vertical plane normal to this minimum horizontal stress, now have no compressional forces holding them closed. From B to C, the stress/strain proportionality no longer exists, such that for unit stress a greater proportion of strain is produced. The pressure difference, C - B, is that pressure necessary to push fluid into the cracks, apply pressure to the walls, and to apply pressure to the leading edge (close

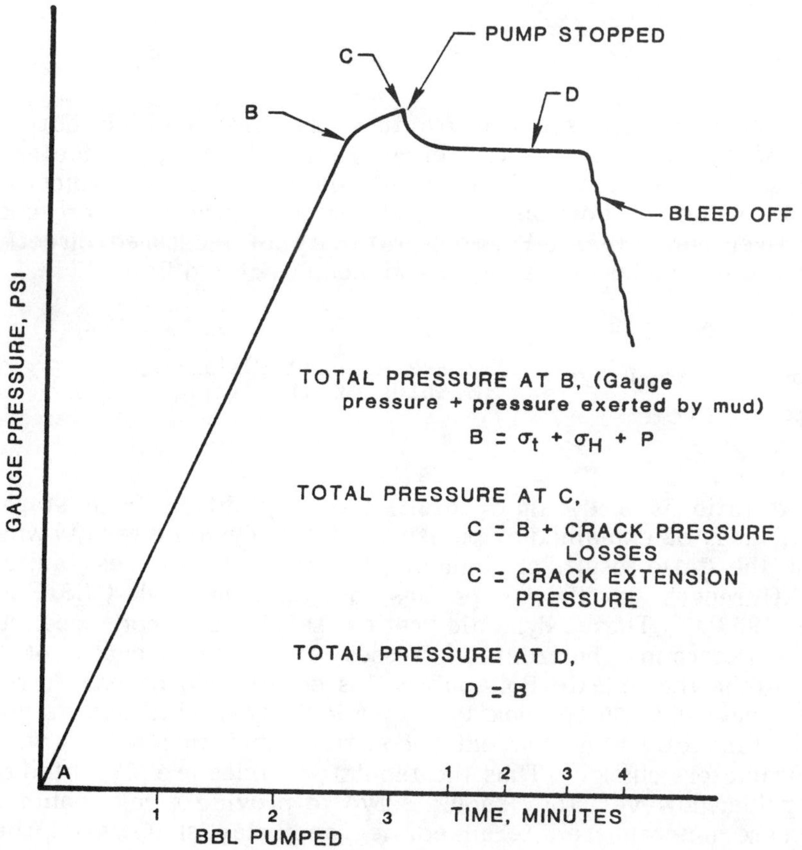


Figure 5-6. Typical fracture-test plot, showing the point at which the minimum horizontal stress becomes balanced by the total pressure within the borehole (B). If B=D, then the volume of mud returned on bleed-off should be equal to the initial volume pumped

to the tip) of the cracks. When the pressure within the borehole is approximately 5 percent greater than the total minimum horizontal stresses, an almost infinite tensile stress occurs at the tips of the crack. At this point, the cracks extend rapidly along the path of minimum resistance, i.e. in a vertical plane, normal to the minimum compressive stress (in a vertical borehole, with horizontal beds). If the pump is stopped at that moment, fracture propagation will cease and the pressure will fall to D. When the pressure in the borehole has fallen (due to the increase in volume caused by the fractures) to a pressure equal to the pore pressure plus the total minimum horizontal stresses, it should stabilize at a pressure equal to B. When the excess pressure is bled off, the amount of returning mud should be almost equal to the amount pumped. If the shut-in pressure (D) is lower than B, then it would be reasonable to assume that the fractures are still open, possibly being propped open by mud contaminants or cuttings. The larger volume produced by the open fractures causes a larger decrease in pressure, such that $B - D > 0$. In this case, the amount of mud returned or bled off is less than the amount pumped. If this occurs in permeable formations, then possibly significant mud losses may occur due to the highly increased surface area in the fractured zone.

## 5.19 METHOD

All the data necessary to estimate fracture pressures can be obtained from interpretation of the first fracture test in compact formation, parameters that are normally measured or calculated when drilling wildcat wells, and typical values for Poisson's ratio. Values of Poisson's ratio, shown in Figure 5-7, were obtained by sonic testing (Weurker, 1963). Poisson's ratio is not measured directly, but is calculated from the modulus of elasticity and modulus of rigidity:

$$\text{Poisson's ratio, } \mu \text{, } = \frac{\text{Modulus of Elasticity}}{2 \text{ (Modulus of Rigidity)}} - 1 \qquad (5\text{-}26)$$

The calculated ratio is a dynamic result and may differ from static elastic properties. This may be explained by pointing out that dynamic results which differ markedly from the static results are indicative of zones of weakness, anisotropy, or directional differences in the properties of the material (U.S. Bureau of Reclamation, 1953). These dynamic ratios should be more realistic when attempting to determine horizontal stresses at depth because of observed anisotropies, rather than static Poisson's ratios determined on carefully selected and prepared specimens. Each rock type (particularly in situ) has its own unique Poisson's ratio (and other mechanical properties), and this will vary when the influencing parameters change. Thus the tabulated values are presented only as an approximate guide; however, they should serve to provide a reasonable estimate. When two or more minerals are intermixed, i.e. sandy clay, shaley sand, the matrix-forming rock type must be determined. If the lithology is a sand with the grains in contact with one another, and clay is the matrix (clay content is less than 30%), the Poisson's ratio is dependent on the sand type. If the clay content is greater than 30%, so that the sand grains are not in contact but are supported in the clay matrix, then Poisson's ratio is dependent on the clay type. Likewise, if a clay is highly calcareous (greater than 50%), the carbonate content may have a significant

| Rock Type | | Poisson's Ratio |
|---|---|---|
| Clay, very wet | | 0.50 |
| Clay | | 0.17 |
| Conglomerate | | 0.20 |
| Dolomite | | 0.21 |
| Greywacke: | coarse | 0.07 |
| | fine | 0.23 |
| | medium | 0.24 |
| Limestone: | fine, micritic | 0.28 |
| | medium, calcarenitic | 0.31 |
| | porous | 0.20 |
| | stylolitic | 0.27 |
| | fossiliferous | 0.09 |
| | bedded fossils | 0.17 |
| | shaley | 0.17 |
| Sandstone: | coarse | 0.05 |
| | coarse, cement | 0.10 |
| | fine | 0.03 |
| | very fine | 0.04 |
| | medium | 0.06 |
| | poorly sorted, clayey | 0.24 |
| | fossiliferous | 0.01 |
| Shale: | calcareous (<50% $CaCO_3$) | 0.14 |
| | dolomitic | 0.28 |
| | siliceous | 0.12 |
| | silty (<70% silt) | 0.17 |
| | sandy (<70% sand) | 0.12 |
| | kerogenaceous | 0.25 |
| Siltstone | | 0.08 |
| Slate | | 0.13 |
| Tuff: | glass | 0.34 |

Figure 5-7. Suggested Poisson's ratios for different lithologies

effect on the mechanical properties, so the Poisson's ratio for shaley limestone should be used. Greater than 80% carbonate content in a shale, or rather 20% clay in a calcareous lithology, indicates that the gradation has progressed essentially from shale to micrite or fine limestone. Careful analysis and interpretation of cuttings and logs should provide a sound basis for selecting the correct Poisson's ratio. The weakest interval in the borehole will be that which has the lowest pore pressure and lowest Poisson's ratio. A low pore pressure in a zone that has a higher Poisson's ratio may have a higher calculated fracture pressure than another zone that has a higher pore pressure and lower Poisson's ratio. Fracture pressures calculated at changes in lithology and pore pressures will show the weakest interval in the borehole.

The result of the first fracture test in compact formations is used to calculate the effective stress ratio of the superposed tectonic stress, if present:

$$\sigma_t = F - \left[\sigma_1' \left(\frac{\mu}{1-\mu}\right)\right] - P \qquad (5\text{-}27)$$

$\sigma_t$ remains directly proportional to $\sigma_1'$, providing the strata remain close to the horizontal and the basin structure does not change significantly with depth.

Since

$$\sigma_t/\sigma_1' = \beta \qquad (5\text{-}20)$$

where $\beta$ defines the stress ratio of $\sigma_t$ to $\sigma_1'$, and remains constant with depth,

then as $\sigma_1'$ is known at any point within the drilled hole,

$$\sigma_1' = S - P \qquad (5\text{-}19)$$

where S and P are the overburden pressure and pore pressure, respectively:

$$\sigma_t = \sigma_1' * \beta \qquad (5\text{-}28)$$

The overburden pressure, S, should be accurately determined from a density log or measured bulk densities for the first fracture pressure test. It is particularly important on offshore wildcats to take into account the air gap and water depth when calculating overburden gradients (Christman, 1973). Pore pressures can be reliably estimated from drilling exponent plots, mud density/gas relationships, and sonic logs.

Accuracy of the parameters when obtaining $\sigma_t$ from the first fracture test is of prime consideration, as any significant error at this point will render false fracture pressures with depth.

Since the local effective stress ratio has now been found, fracture pressures can be calculated as the well progresses, as changes in lithology (Poisson's ratio), pore pressure, and overburden pressure occur:

$$F = \sigma_t + \sigma_1' \left(\frac{\mu}{1-\mu}\right) + P \qquad (5\text{-}25)$$

Between log runs the overburden gradient may be extrapolated with a reasonable degree of accuracy by plotting overburden pressure with depth (Figure 5-8). It will be seen that the relation is approximately linear, except for the upper portion of the curve which is affected by water depth, uncompacted sediments and the air gap. Linear extrapolation of the trend may be achieved with confidence, providing the upper overburden gradient obtained from logs or bulk densities was accurate. Correction of the extrapolated trend may be necessary after subsequent logging runs, or continuously updated from bulk density measurements.

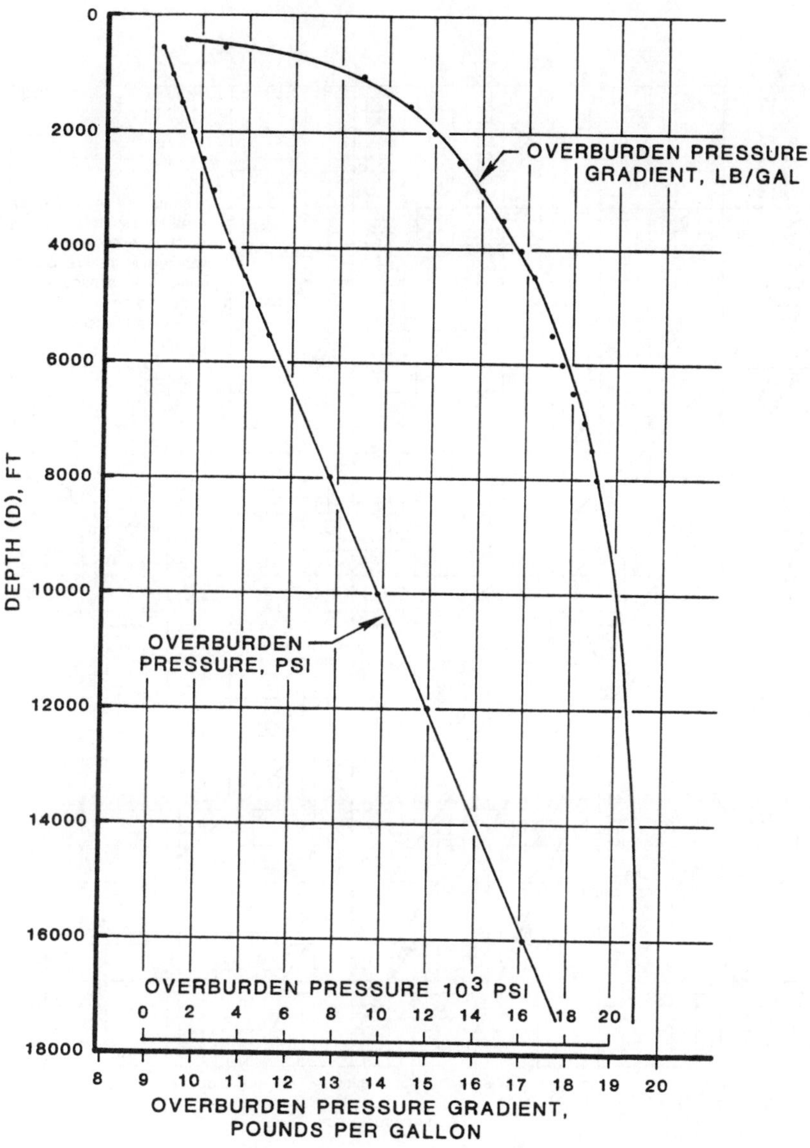


Figure 5-8. Typical overburden curve from an offshore well

A continuous, real-time plot of calculated fracture pressures with depth is thus made possible, providing the various Poisson's ratios can be adequately determined from the cuttings. If complex or interrelated lithologies are encountered, assignment of a unique Poisson's ratio may not be immediately apparent: of the several lithologies that may occur in the same sample, that which has the lowest Poisson's ratio should be used until confirmation is obtained from logs. If the pore pressure gradient remains constant with depth, then the $\sigma_1'$, $\sigma_t$ and $\sigma_H$ (with constant lithology) gradients are constant (Figure 5-9). Fluctuating pore pressure causes significant changes in all the stress gradients (Figure 5-10).

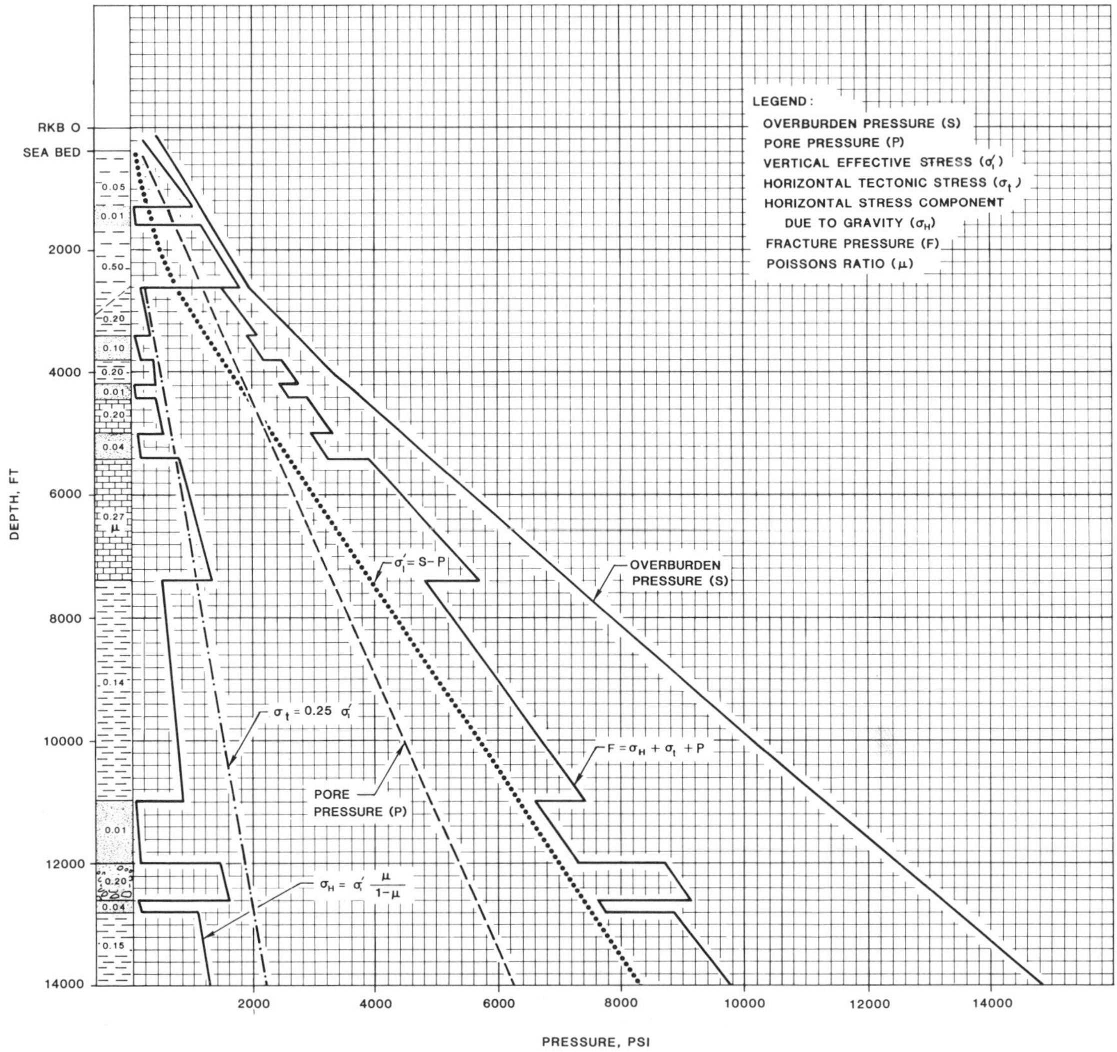


Figure 5-9. Hypothetical changes in $\sigma_1$, $\sigma_t$ and $\sigma_H$ with depth and constant pore pressure gradient. Resultant fracture pressure curve is shown.

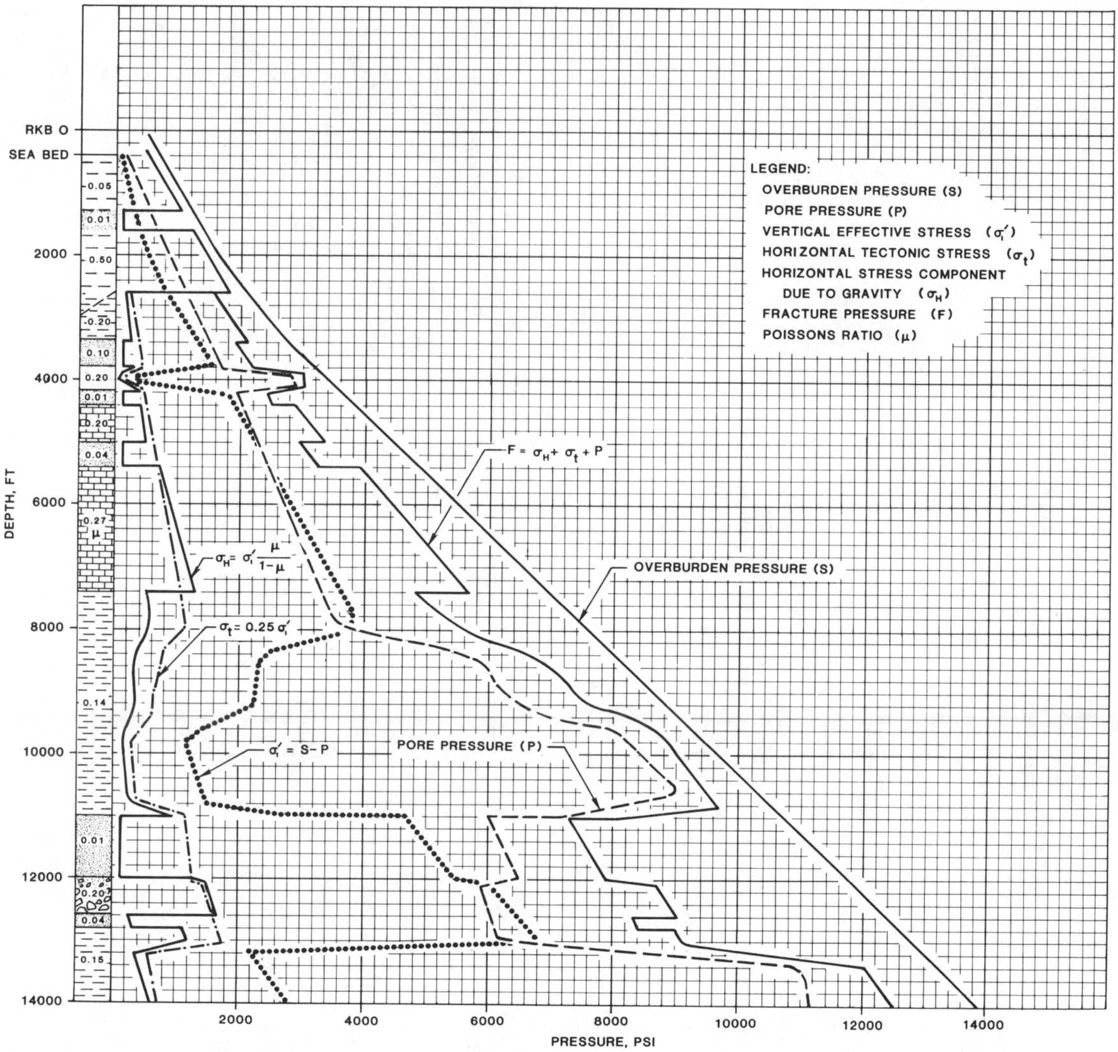


Figure 5-10. Hypothetical changes in $\sigma_1$, $\sigma_t$ and $\sigma_H$ with depth and changing pore pressure gradient. Note that all stresses are affected by the pore pressure. Resultant fracture pressure curve is shown.

IRWELL VALLEY DRILLING CONSORTIUM : TEST WELL # 1
Time : 3:53 Date : 5/ 9/81

FRACTURE GRADIENT CALCULATIONS

Input Data:

| | | |
|---|---|---|
| Vertical Depth of Fracture Test | 8000.0 ft | |
| For Weakest Formation: | | |
| Vertical Depth | 8000.0 ft | |
| Poisson's Ratio | .05 | |
| Estimated Pore Pressure | 9.50 lb/gal | 3944 psi |
| Mud Hydrostatic Pressure | 12.00 lb/gal | 4982 psi |
| Gauge Pressure at Leak Off | .72 lb/gal | 300 psi |
| Total Fracture Pressure | 12.72 lb/gal | 5282 psi |
| Overburden Pressure | 16.50 lb/gal | 6851 psi |

Calculated Data:

| | |
|---|---|
| Effective Vertical Stress | 2906 psi |
| Effective Horizontal Stress | 153 psi |
| Tectonic Stress | 1185 psi |
| Effective Stress Ratio (Beta) | .408 |

| VERTICAL DEPTH ft | PORE PRESSURE Gradient lb/gal | PORE PRESSURE Gradient psi/ft | PORE PRESSURE Press. psi | OVERBURDEN Gradient lb/gal | OVERBURDEN Gradient psi/ft | OVERBURDEN Press. psi | POISSON'S RATIO u | EFFECTIVE VERTICAL STRESS psi | SUPERPOSED TECTONIC STRESS psi | HORIZONTAL STRESS DUE TO GRAVITY psi | FRACTURE PRESSURE Gradient lb/gal | FRACTURE PRESSURE Gradient psi/ft | FRACTURE PRESSURE Press. psi |
|---|---|---|---|---|---|---|---|---|---|---|---|---|---|
| 9000.0 | 10.20 | .529 | 4764 | 18.00 | .934 | 8408 | .10 | 3643 | 1486 | 405 | 14.25 | .739 | 6655 |
| 9100.0 | 10.25 | .532 | 4841 | 18.05 | .937 | 8525 | .20 | 3684 | 1502 | 921 | 15.38 | .798 | 7264 |
| 9200.0 | 10.27 | .533 | 4904 | 18.05 | .937 | 8619 | .16 | 3715 | 1515 | 708 | 14.92 | .775 | 7126 |
| 9300.0 | 10.28 | .534 | 4962 | 18.10 | .939 | 8736 | .05 | 3774 | 1539 | 199 | 13.88 | .720 | 6699 |
| 9400.0 | 10.30 | .535 | 5025 | 18.10 | .939 | 8830 | .12 | 3805 | 1552 | 519 | 14.54 | .755 | 7095 |
| 9500.0 | 10.30 | .535 | 5078 | 18.15 | .942 | 8949 | .22 | 3870 | 1578 | 1092 | 15.71 | .816 | 7748 |
| 9600.0 | 10.35 | .537 | 5157 | 18.20 | .945 | 9068 | .25 | 3911 | 1595 | 1304 | 16.17 | .839 | 8055 |
| 9700.0 | 10.40 | .540 | 5236 | 18.20 | .945 | 9162 | .10 | 3927 | 1601 | 436 | 14.45 | .750 | 7273 |
| 9800.0 | 11.00 | .571 | 5595 | 18.25 | .947 | 9282 | .10 | 3687 | 1504 | 410 | 14.76 | .766 | 7508 |
| 9900.0 | 11.10 | .576 | 5703 | 18.30 | .950 | 9403 | .27 | 3699 | 1508 | 1368 | 16.70 | .867 | 8580 |

Figure 5-11. ELOS fracture gradient calculation

Figure 5-11 is an example of the ELOS fracture gradient calculation.

A problem that may be encountered in using this method in the field is with drilling personnel who have become familiar with the Eaton method and to using the empirical, Gulf Coast, variable overburden "Poisson's ratios" (Figure 5-4). It will be necessary to account to such personnel for the difference in the value of Poisson's ratio used in the Exploration Logging method. This may be done by substituting Equations (5-19) and (5-28) into Equation (5-25) and dividing by vertical depth to obtain gradients, thus obtaining the equation in the form

$$\frac{F}{D} = \left(\frac{S}{D} - \frac{P}{D}\right) * \left[\left(\frac{\mu}{1-\mu}\right) + \beta\right] + \frac{P}{D} \qquad (5\text{-}29)$$

This is directly comparable to Eaton's method

$$\frac{F}{D} = \left(\frac{S}{D} - \frac{P}{D}\right)\left(\frac{\mu_e}{1-\mu_e}\right) + \frac{P}{D} \qquad (5\text{-}5)$$

where

$\mu$ = true Poisson's ratio

$\mu_e$ = Eaton's empirically derived Poisson's ratio

It is obvious that these two quantities are unlike and cannot be used interchangeably. The Eaton Poisson's ratio will be a function of the true Poisson's ratio and the regional stress ratio

$$\mu_e = \frac{\left[\left(\frac{\mu}{1-\mu}\right) + \beta\right]}{1 + \left[\left(\frac{\mu}{1-\mu}\right) + \beta\right]} \qquad (5\text{-}30)$$

When applying the Exlog (Zero Tensile Strength) method you must make sure that the client is familiar with its derivation. It must be explained that the method uses Poisson's ratio values dependent only upon lithology and a regional stress ratio determined on and for that particular well and basin. Unlike the empirically derived Eaton quantity, the true Poisson's ratio for a particular lithology does not include a regional stress component and will not vary with depth or between basins.

Several factors affect fracture test pressures, aside from formation characteristics:

1. Higher mud densities appear to cause higher fracture pressures (MacPherson and Berry, 1972), although this may be due to a related increase in viscosity.

2. Smaller hole diameters may cause higher fracture pressures (Haimson and Fairhurst, 1969).

3. The rate of pressurization affects fracture pressures: high pump rates produce inflated fracture pressures (Haimson and Fairhurst, 1969). This effect is smaller than that in (2) above.

4. High mud gel strengths require higher pressures to initiate circulation. Correction for this pressure loss can be obtained from Chenevert and McClure, 1978.

5. Hole deviation significantly affects fracture pressures (Bradley, 1979).

6. Rig and sensor instrumentation probably are accurate to within 5% (Taylor and Smith, 1970). Accuracy of predicted fracture pressures is therefore limited to this range.

7. Mud penetrability does not alter the actual breakdown pressure, but it will affect the shape of the fracture pressure plot such that the point at which the total horizontal minimum stress is balanced may be obscured.

A combination of these mechanisms is probably responsible for a considerable scatter of data points. However, if fracture test procedures are kept as consistent as possible on any one well, then the results obtained should lie within the 5% instrument error margin.

## 5.20 SUMMARY

A theoretical model is put forward that attempts to describe the principal stress system within a basin of simple topography and structure. If a well is drilled nearly vertically, then the well should be approximately parallel to one of the principal stresses, which is equal to the effective pressure of the overlying strata. The horizontal stresses are a combination of the stress caused by gravity and a superposed horizontal tectonic stress. The latter may be nonexistent or may reach a maximum of two to three times the vertical stress (Hubbert and Willis, 1957). The minimum horizontal stress is measured by the first fracture test in compact formation. As the vertical stress increases approximately linearly with depth, then the tectonic horizontal stress will increase linearly with depth also, defined by a constant stress ratio, $\beta$. Since this ratio is obtained from the first fracture test, then at any subsequent depths the fracture pressures may be calculated providing pore pressures, overburden pressures and lithological relationships are known. The following may be concluded:

1. Fracture pressures may be estimated when drilling rank wildcat wells to within an error margin of approximately 5%.

2. Fracture pressures are dependent on the total minimum horizontal stress (a combination of a stress caused by gravity and a superposed tectonic stress) and the pore pressure.

3. Factors affecting actual fracture pressures may be minimized by conducting fracture tests as consistently as possible. A correction is available for gel strength (usually $<$ 0.1 lb/gal), but changes in mud types or large changes in properties may cause significant deviation from calculated fracture pressures. It is also suggested that at least one circulation is effected prior to conducting a fracture test, in order to minimize any inconsistencies in the mud column.

4. The theoretical fracture pressure formula provides an explanation for fracture pressures that equal the overburden pressure in shallow wet clays, and also indicates that if a sandstone reservoir is fractured, the fracture should not extend into or through the seal. Probably an inherent property of a permeability seal is a relatively high Poisson's ratio: these rock types require a higher pressure within the borehole to balance the horizontal compressive stress, so a hydraulic fracture within an underlying permeable stratum should be confined to that stratum.

## 5.21 EXAMPLES

1. A 12-1/4-inch pilot hole has been drilled to 1500 feet offshore. Water depth is 200 feet, and RKB to sea level is 100 feet. The entire sequence is soft, unconsolidated clays. The hole is opened to 26 inches, and 20-inch casing is run and cemented at 1460 feet. After drilling out the shoe, the rat hole is cleaned, a full mud circulation is allowed before pulling the bit up into the casing shoe (closing the annular preventer), and a fracture test is performed.

   Fracture occurred at 14.3 lb/gal EMD. Analysis of the data indicated that the result is normal: although the formation balance gradient was normal at 8.6 lb/gal and the calculated overburden gradient at 1460 ft was 14.1 lb/gal, Poisson's ratio for the wet clay would be close to 0.5, hence

$$F = \sigma_t + \sigma_1' \left(\frac{\mu}{1-\mu}\right) + P \qquad (5\text{-}25)$$

   where

   $\sigma_t$ = 0; the rock is effectively water-supported

   $\sigma_1'$ = 14.1 - 8.6 = 5.5

   $\mu$ = 0.5

   P = 8.6

   predicted F = 14.1 lb/gal. $\left(0 + 5.5 * \frac{0.5}{0.5} + 8.6 = 14.1\right)$

   The fracture would be horizontal. Note that this example cannot be used to predict further fracture tests with depth, as $\sigma_t$ is nonexistent due to the fact that the wet, unconsolidated clay has negligible shear strength and thus could not support an applied tectonic stress.

2. At 3300 feet, 13-3/8-inch casing is run to 3270 feet. The formation balance gradient is normal (8.6 lb/gal), the estimated overburden gradient at 3270 feet is 16.4 lb/gal, and the lithology in 30 feet of open hole is clay with a sandstone bed at 3290 feet. The sandstone is coarse grained and well sorted. Figure 5-7 gives a Poisson's ratio of 0.05 for this sand. Assuming no tectonic stress, i.e. $\sigma_t$ = 0, then the predicted fracture pressure would be

$$F_{calc} = 0 + 1332\left(\frac{0.05}{1 - 0.05}\right) + 1469 \text{ psi}$$

$$F_{calc} = 1539 \text{ psi}$$

$$= 9.0 \text{ lb/gal}$$

The actual fracture pressure was 1911 psi, indicating that a tectonic stress is present. $\sigma_t$ is thus found from:

$$\sigma_t = F - F_{calc}$$

$$= 1911 - 1539$$

$$\sigma_t = 372 \text{ psi}$$

This value, $\sigma_t$ = 372 psi, indicates that a considerable tectonic stress is apparent. In order to estimate further fracture pressure with depth, the $\sigma_t/\sigma_1'$ ratio has to be found:

$$\beta = \frac{\sigma_t}{\sigma_1'} \qquad (5\text{-}20)$$

$$= \frac{372}{1332}$$

$$= 0.279$$

3. Utilizing $\beta$, pore pressure estimations, estimated overburden pressures, and Poisson's ratios for subsequent lithologies, the fracture pressures may now be estimated at any point. The tectonic stress at any depth can be found by:

$$\sigma_t = \sigma_1' * \beta \text{ , psi} \qquad (5\text{-}28)$$

## 5.22 MASSIVE HYDRAULIC FRACTURING (MHF) AND STIMULATION

Stimulation of a well is undertaken when the ability of the rock to allow the passage of fluids is increased (1) by the creation of fractures, (2) by enlarging existing openings in the rocks adjacent to the wells, or (3) by removing deposits that have partially blocked the openings during earlier production. Fractures may be created or widened by hydraulic fracturing and kept open by injecting a suitable proppant held in suspension by a viscous gel. The gel is later recovered, leaving a permeable conduit from the formation to the wellbore. Explosives may also be used to create fractures.

The intervals to be fractured are isolated by removable packers, and usually a low-viscosity, highly penetrating fluid is injected to create initial fractures. This fluid is then rapidly followed by a large volume of gel containing suitable proppant (usually very well rounded, well-sorted sand, or pellets of high crushing strength). The height and length of the created fractures can largely be controlled by the rate and volume of material pumped into the formation. Common fracture dimensions (calculated) for a 100-ft reservoir unit would be a fracture of approximately 400 feet in length and 80 to 100 feet in vertical extent.

Acid may be injected to widen openings by solution of the rock, while organic solvents may be used to remove clogging waxy and asphaltic deposits, or to remove filtrate and mud invasion from the borehole wall. Occasionally, the reservoir permeability adjacent to the well is severely impaired during drilling due to excessive overbalance. Fractures induced in these zones will greatly improve flow without the need to restore the damaged zone to its undamaged condition.

## 5.23 KICKS AND KICK TOLERANCE

A kick is a problem that does not commonly occur. However, the penalty for failure to control the well is the loss of the well, and occasionally the loss of the rig and the lives of the crew. Unreasonable procedures can in themselves cause hazardous conditions that severely jeopardize safety. Blowouts are a disaster from the viewpoint of people, economics, politics, and the environment.

A standard kick control procedure may vary from rig to rig, but generally four simultaneous operations are considered.

- Rig Control. This includes the blowout preventers, pumps, drawworks, and other rig operating equipment that is necessary. Rig control is the responsibility of the driller, and any blowout control procedure should assign these operations to the driller.

- Mud Control. This involves adding barite for increasing the mud density, but also includes adding chemicals to the drilling mud and proper operation of the mixing systems. The mud control operations are generally the responsibility of the derrick man and mud engineer.

- Choke Control. This includes calculating the proper pressures and time relationships as well as correctly operating the choke and monitoring the pump rate. The choke operator should be the best trained man on the rig from the viewpoint of kick control. He is required to give procedural guidance during the well killing operation.

- Supervision. This is the final element of control during a well kick. The tool pusher is the normal rig and crew supervisor and this should be his task. To assign the job of choke operator to the tool pusher is undesirable because he would then be restricted to the rig floor. The rig, during the critical well control procedure, needs a general overall supervisor, and this job is best undertaken by the tool pusher who knows both the rig and the crew.

Decisions made under pressure depend completely on the knowledge, attitude, and judgement of the supervisor. They can be confused by crew change problems and divided responsibilities between the tool pusher and drilling foreman, or drilling engineer. So one of the most important elements of a kick control package is the establishment of a policy and prodecure outlined in whatever degree of detail necessary. These procedures must be known by all members of the Exlog crew: it is the responsibility of the P.E.G. or GEMDAS Trailer Captain to obtain this information.

## 5.24 CAUSES OF KICKS

There are five major causes of well kicks:

1. A failure to keep the hole full. The majority of blowouts occur when the bit is off-bottom when tripping. When the pumps are shut down prior to tripping, there is a pressure reduction in the borehole equal to the annular pressure losses. If the mud pressure and pore pressure are almost equal, flow may occur when circulation stops. As pipe is removed, the mud level in the hole falls, causing a pressure reduction in open hole. The pipe displacement must be converted to pump strokes so that it will be known precisely how many strokes are necessary to fill the hole.

2. Swabbing. When pipe is pulled it acts as a piston, more so below than above the bit. Gel strength and viscosity of the mud have a large effect on swabbing. Swabbing is further increased if mud cake is thick, the bit is balled-up and the jets are blocked, or if there is a backpressure valve in the string. The speed at which pipe is pulled has a great effect on hole swabbing. On GEMDAS units, the EAP Swab/Surge program provides a range of pipe-pulling speeds and the corresponding swab and surge pressures. (Figure 5-12 is an example printout of the Swab/Surge program.) If swabbing does occur, pipe should be run back to bottom and the invading fluid should be circulated out. Surge pressures, when running into the hole (with pipe or casing), may be sufficient to overcome the fracture pressure of a weak formation. Again, the Swab/Surge printout should be consulted and the pipe should be run at a speed that produces surge pressures below the minimum fracture pressure. (This is important when running in pipe or casing anywhere in the hole, as pressures are transmitted to open hole even when the bit or casing shoe is still inside the previous casing.)

3. Insufficient mud density. Fewer kicks result from too low a mud density than from the above two causes. If a kick occurs while drilling due to insufficient mud density, there is a possibility that an oversight has occurred — or equally probable is the possibility of poor engineering practice. In any event, trends and plots will have to be reevaluated. Penetration of a geopressured formation without prior indication may have occurred — for example, by crossing a fault or unconformity. On the other hand, changes in lithology or drilling practices may have masked a transition zone.

4. Poor well planning. The mud and casing programs have a great bearing on kick control. These programs must be flexible so that progressively deeper casing strings can be set; otherwise, a situation may arise where it is not

IRWELL VALLEY DRILLING CONSORTIUM : TEST WELL # 1
3:43 5/ 9/81

SWAB AND SURGE ANALYSIS

TOTAL DEPTH (ft) = 15000.0, CASING SHOE DEPTH =8000.0, OPEN PIPE
MUD WEIGHT = 12.00 lb/gal, PLASTIC VISCOSITY = 28.00, YIELD POINT = 14.00
LOW RANGE POWER LAW: K = .452, N = .737
MID RANGE POWER LAW: K = .452, N = .737

| RUNNING SPEED | | BIT ON BOTTOM | | | BIT AT SHOE | | |
|---|---|---|---|---|---|---|---|
| ft/mi | SEC/STND | psi | TD EQUIV MUD WT | | psi | SHOE EQUIV MUD WT | |
| | | | SURGE | SWAB | | SURGE | SWAB |
| 558.00 | 10 | 465.9 | 12.63 | 11.37 | 289.6 | 12.70 | 11.30 |
| 279.00 | 20 | 118.9 | 12.16 | 11.84 | 78.1 | 12.19 | 11.81 |
| 186.00 | 30 | 68.7 | 12.09 | 11.91 | 40.4 | 12.10 | 11.90 |
| 139.50 | 40 | 55.2 | 12.07 | 11.93 | 32.3 | 12.08 | 11.92 |
| 111.60 | 50 | 46.8 | 12.06 | 11.94 | 27.4 | 12.07 | 11.93 |
| 93.00 | 60 | 40.9 | 12.06 | 11.94 | 24.0 | 12.06 | 11.94 |
| 79.71 | 70 | 36.5 | 12.05 | 11.95 | 21.4 | 12.05 | 11.95 |
| 69.75 | 80 | 33.1 | 12.04 | 11.96 | 19.4 | 12.05 | 11.95 |
| 62.00 | 90 | 30.3 | 12.04 | 11.96 | 17.8 | 12.04 | 11.96 |
| 55.80 | 100 | 28.1 | 12.04 | 11.96 | 16.4 | 12.04 | 11.96 |
| 50.73 | 110 | 26.2 | 12.04 | 11.96 | 15.3 | 12.04 | 11.96 |
| 46.50 | 120 | 24.5 | 12.03 | 11.97 | 14.4 | 12.03 | 11.97 |
| 42.92 | 130 | 23.1 | 12.03 | 11.97 | 13.6 | 12.03 | 11.97 |
| 39.86 | 140 | 21.9 | 12.03 | 11.97 | 12.8 | 12.03 | 11.97 |
| 37.20 | 150 | 20.8 | 12.03 | 11.97 | 12.2 | 12.03 | 11.97 |
| 34.88 | 160 | 19.9 | 12.03 | 11.97 | 11.6 | 12.03 | 11.97 |
| 32.82 | 170 | 19.0 | 12.03 | 11.97 | 11.1 | 12.03 | 11.97 |
| 31.00 | 180 | 18.2 | 12.02 | 11.98 | 10.7 | 12.03 | 11.97 |
| 29.37 | 190 | 17.5 | 12.02 | 11.98 | 10.2 | 12.02 | 11.98 |
| 27.90 | 200 | 16.8 | 12.02 | 11.98 | 9.9 | 12.02 | 11.98 |

Figure 5-12. Typical Swab/Surge printout from a 15,000-ft well with a a mud density of 16.5 lb/gal, a PV of 35 and a YP of 20

possible to control kicks or lost circulation. Kick control is an important part of a complete well plan, but the importance of kick prevention should not be overstated to the point that overall drilling effectiveness is reduced.

5. Lost circulation. Raising the mudweight to a value that exceeds the lowest fracture pressure for fear of a kick is not as commonplace today as it was in the 1940s and 1950s. A kick may still occur, but it is more often due to the

fracturing of a formation at a lower pore pressure than to fracturing an abnormal pressure zone above. Rather than setting another casing string after drilling through an abnormally high pore pressure zone, mudweight is kept high to balance these fluids so that, if subsequently the pore pressure decreases significantly, the lower formations become susceptible to fracturing. If fracturing does occur, the mud level in the annulus drops due to lost circulation, the reduction in mud pressure may allow influx of formation fluids, and a well kick. The existence of an abnormal pressure zone and a lost circulation zone in any hole section are two ingredients necessary for the kick recipe, and utmost care combined with diligent observation is necessary to drill such a section successfully.

## 5.25 RECOGNITION OF KICKS

The only time a blowout can occur totally without warning is when a well is drilled offshore and there is no annular connection from the wellhead to the rig. However, there is never lack of indications that a kick or blowout may occur. Since in the majority of cases the borehole and mud pits are a closed circulating system, the addition of any fluid from the formation will be displayed as a change in flowrate and a change in total mud volume.

One (fortunately rare) occurrence when surface recognition may be delayed is when, due to fracturing and lost circulation, the borehole is not filled with mud and cannot be filled. That is, the rate of loss is greater than the rate at which mud can be pumped into the hole. In such circumstances it is not possible to monitor mud level in the hole. A major influx into the borehole may occur and will go undetected at surface. To prevent this possibility the well should be shut in, shut-in pressures (if any) carefully monitored, pipe movement carried out by "stripping" through the blowout preventors, and hole fill-up by the kill and choke lines.

### 5.26 During Connections

When drilling with a close balance between borehole pressure and pore pressure, flow into the wellbore can occur when the pumps are shut down for a connection. This results from a pressure reduction because of the reversal in value of the annular pressure loss. When pumping, the annular pressure loss adds to the static mud pressure, so when the pumps are shut down this extra pressure is lost; and when the kelly is lifted, swab pressures further reduce the bottomhole pressure. (Refer to Section 3 for ECD calculation.) An increase in hookload may indicate that a lighter fluid has invaded the hole. In a less dense fluid the buoyancy will be less and the string will have greater weight.

### 5.27 While Tripping

Since kick control procedure is greatly simplified when the bit is near the bottom of the hole, kicks during a trip are the greatest potential danger. With the pipe out of the hole, it is impossible to get heavier mud to the bottom. During a trip, the same reversal of the annular pressure drop occurs as described in paragraph

5.26. However, in addition, because pipe is being removed, the hole must regularly be topped up with mud — usually every five to ten stands. If the hole does not take enough mud to displace the volume of withdrawn drillpipe, this is an indication that pore fluid is displacing the mud near the bottom of the hole and that the well is kicking. While tripping out on modern offshore rigs, the hole is kept full by continually circulating through a tall, narrow trip tank. By closely monitoring the decrease in tank volume against the calculated volume, any discrepancy can be noticed almost immediately. In GEMDAS units, the Trip Monitor program provides comparison of volumes for every stand pulled. (An example Trip Monitor program printout is shown in Figure 5-13.) Alternatively, the Trip Condition Log provides a summary of hole condition and fill-up during trips (Figure 5-14). Older offshore rigs and land rigs may not have a trip tank, so reliance for volume checks is placed on monitoring one active pit and the pump strokes. Pit Volume Totalizer monitoring provides a cross-check, but, because of the large surface area of the pits, precision may be limited. The mud pump is a reasonably efficient displacement monitor at low pressures and rates, and pump strokes are often used to measure the proper amount of displaced fluid.

It is normal for the hole to take slightly more mud than the volume of the pipe removed, due to static filtration into the formation. If a kick occurs when the bit is not on bottom, every effort must be made to run back in the hole. Modern BOP stacks are designed for reliable pipe stripping through the bag-type or ram sets, enabling in most cases the bit to be run to TD.

## 5.28 Sequence of Events

In most cases, the following distinct series of events may lead up to a kick when drilling. Some indications may not occur, however, whereas others may be accentuated. Recognition of the changing trends at an early stage should allow remedial action to be taken, thus minimizing the inherent potential hazards and costs.

1. The first indication that a kick may occur is a drilling break. The fast drilling break need not necessarily indicate an increase in porosity, permeability and pressure, but it is prudent to assume that it does. The magnitude of the drilling break varies from a slight increase, about 50 percent, to 200 or 300 percent in soft sediments. Any significant drilling break must be checked for flow. This is done by (1) picking up the kelly so that the bushings are about 10 feet above the rig floor, (2) stopping the pumps, and (3) observing the level in the annulus with a torch (flashlight) to see whether flow is occurring. This may be difficult on floating rigs, as the level fluctuates with the heave of the vessel. In these instances the flowcheck should last for at least 5 minutes so that a flow may be verified. If the well does flow, it should be shut in and checked for pressures.

2. The second indication of a kick, or first confirmation that a kick is occurring, is an increase in flowrate in the flowline. The entrance of any formation fluid into the wellbore causes the flowrate to increase, and this occurs concurrently with or shortly after the drilling break. The intruding

UNIT #155 10:13 7/20/78

TRIP OUT, OPEN PIPE, BIT NO: 9

HOOKLOAD: MAX(+), EXPECT(*)

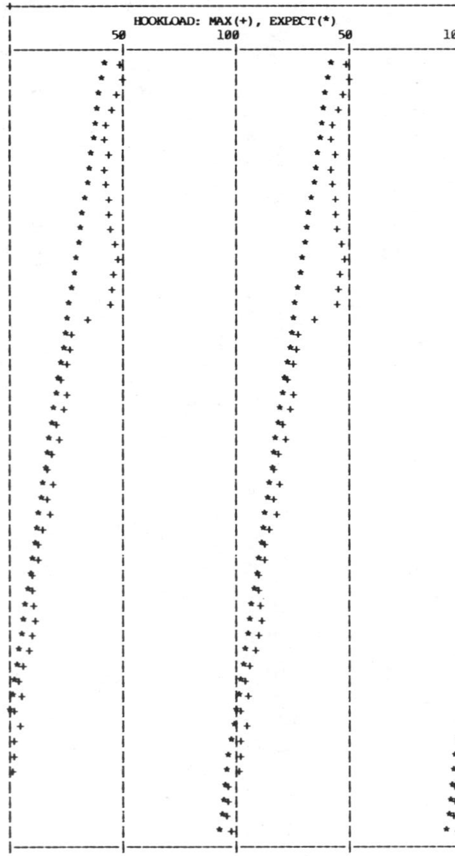

| TIME | DEPTH BIT | HOLE SIZE | STANDS DONE | STANDS TO GO | HOOKLOAD MAX | HOOKLOAD EXP | HOOKLOAD DIF | PIT VOL ACT | PIT VOL EXP | PIT VOL DIF | CMPL TIME | AVE ft/min | SEC/STND | SWAB PR | FILE NO. |
|---|---|---|---|---|---|---|---|---|---|---|---|---|---|---|---|
| 1017 | 8451 | 8.500 | 1 | 93 | 249.0 | 241.7 | 7.31 | 464 | 462 | 2 | | 150.4 | 36 | 75.9 | 2.1 T |
| 1020 | 8360 | 8.500 | 2 | 92 | 250.4 | 240.8 | 9.58 | 464 | 462 | 2 | | 129.3 | 42 | 74.4 | 3.1 T |
| 1023 | 8269 | 8.500 | 3 | 91 | 248.0 | 239.8 | 8.14 | 460 | 461 | -1 | | 140.2 | 39 | 75.3 | 4.1 T |
| 1026 | 8178 | 8.500 | 4 | 90 | 245.7 | 238.9 | 6.87 | 461 | 461 | -0 | | 135.6 | 40 | 74.9 | 5.1 T |
| 1029 | 8087 | 8.500 | 5 | 89 | 243.1 | 237.9 | 5.20 | 458 | 460 | -2 | | 142.2 | 38 | 75.4 | 6.1 T |
| 1032 | 7996 | 8.500 | 6 | 88 | 242.2 | 236.9 | 5.31 | 461 | 460 | 1 | | 130.5 | 42 | 74.5 | 7.1 T |
| 1035 | 7905 | 8.500 | 7 | 87 | 244.0 | 236.0 | 8.03 | 457 | 459 | -2 | | 133.3 | 41 | 74.8 | 8.1 T |
| 1038 | 7814 | 8.500 | 8 | 86 | 242.2 | 235.0 | 7.18 | 457 | 459 | -2 | | 136.6 | 40 | 77.0 | 9.1 T |
| 1041 | 7723 | 8.500 | 9 | 85 | 242.7 | 234.1 | 8.64 | 458 | 458 | -1 | | 137.5 | 40 | 77.2 | -10.1 T |
| 1044 | 7632 | 8.500 | 10 | 84 | 244.1 | 233.1 | 11.00 | 458 | 458 | 0 | | 124.6 | 44 | 73.0 | 11.1 T |
| 1047 | 7541 | 8.500 | 11 | 83 | 244.5 | 232.1 | 12.36 | 457 | 457 | 0 | | 153.3 | 36 | 76.1 | 12.1 T |
| 1050 | 7450 | 8.500 | 12 | 82 | 244.8 | 231.2 | 13.62 | 455 | 457 | -2 | | 133.9 | 41 | 74.8 | 13.1 T |
| 1053 | 7359 | 8.500 | 13 | 81 | 246.9 | 230.2 | 16.64 | 456 | 456 | -1 | | 145.9 | 37 | 75.6 | 14.1 T |
| 1056 | 7268 | 8.500 | 14 | 80 | 248.2 | 229.3 | 18.98 | 454 | 456 | -2 | | 120.5 | 45 | 73.7 | 15.1 T |
| 1059 | 7177 | 8.500 | 15 | 79 | 246.1 | 228.3 | 17.83 | 456 | 455 | 1 | | 133.5 | 41 | 74.8 | 16.1 T |
| 1102 | 7086 | 8.500 | 16 | 78 | 245.4 | 227.3 | 18.02 | 455 | 455 | -0 | | 136.5 | 40 | 75.0 | 17.1 T |
| 1105 | 6995 | 8.500 | 17 | 77 | 245.1 | 226.4 | 18.68 | 456 | 454 | 2 | | 138.8 | 39 | 75.2 | 18.1 T |
| 1108 | 6904 | 8.500 | 18 | 76 | 234.8 | 225.4 | 9.38 | 454 | 454 | -0 | | 123.7 | 44 | 74.0 | 19.1 T |
| 1111 | 6813 | 8.500 | 19 | 75 | 227.7 | 224.5 | 3.22 | 451 | 453 | -2 | | 124.1 | 44 | 74.0 | 20.1 T |
| 1114 | 6722 | 8.500 | 20 | 74 | 226.8 | 223.5 | 3.28 | 453 | 453 | 1 | | 120.5 | 45 | 73.7 | 21.1 T |
| 1117 | 6631 | 8.500 | 21 | 73 | 225.8 | 222.5 | 3.25 | 453 | 452 | 1 | | 127.1 | 43 | 74.3 | 22.1 T |
| 1120 | 6540 | 8.500 | 22 | 72 | 223.1 | 221.6 | 1.56 | 454 | 452 | 2 | | 143.7 | 38 | 75.5 | 23.1 T |
| 1123 | 6449 | 8.500 | 23 | 71 | 225.9 | 220.6 | 5.30 | 453 | 451 | 2 | | 128.3 | 43 | 74.4 | 24.1 T |
| 1126 | 6358 | 8.500 | 24 | 70 | 224.6 | 219.7 | 4.90 | 450 | 451 | -0 | | 123.5 | 44 | 73.9 | 25.1 T |
| 1129 | 6267 | 8.500 | 25 | 69 | 221.3 | 218.7 | 2.63 | 449 | 450 | -1 | | 118.7 | 46 | 73.5 | 26.1 T |
| 1132 | 6176 | 8.500 | 26 | 68 | 222.5 | 217.7 | 4.71 | 450 | 450 | 1 | | 124.1 | 44 | 74.0 | 27.1 T |
| 1135 | 6085 | 8.500 | 27 | 67 | 219.1 | 216.8 | 2.29 | 448 | 449 | -1 | | 123.9 | 44 | 74.0 | 28.1 T |
| 1137 | 5994 | 8.840 | 28 | 66 | 217.0 | 215.8 | 1.14 | 450 | 449 | 1 | | 154.3 | 35 | 90.2 | 29.1 T |
| 1139 | 5903 | 8.840 | 29 | 65 | 219.9 | 214.9 | 5.06 | 450 | 448 | 2 | | 149.3 | 37 | 89.9 | 30.1 T |
| 1141 | 5812 | 8.840 | 30 | 64 | 217.0 | 213.9 | 3.10 | 450 | 448 | 2 | | 176.9 | 31 | 91.3 | 31.1 T |
| 1143 | 5721 | 8.840 | 31 | 63 | 218.4 | 212.9 | 5.46 | 447 | 447 | -0 | | 146.6 | 37 | 89.7 | 32.1 T |
| 1145 | 5630 | 8.840 | 32 | 62 | 215.2 | 212.0 | 3.23 | 445 | 447 | -2 | | 160.7 | 34 | 90.5 | 33.1 T |
| 1148 | 5539 | 8.840 | 33 | 61 | 212.8 | 211.0 | 1.78 | 447 | 446 | 1 | | 135.3 | 40 | 88.9 | 34.1 T |
| 1150 | 5448 | 8.840 | 34 | 60 | 213.0 | 210.1 | 2.96 | 448 | 446 | 2 | | 159.4 | 34 | 90.4 | 35.1 T |
| 1152 | 5357 | 8.840 | 35 | 59 | 210.2 | 209.1 | 1.05 | 447 | 445 | 2 | | 146.7 | 37 | 89.7 | 36.1 T |
| 1154 | 5266 | 8.840 | 36 | 58 | 210.6 | 208.1 | 2.44 | 445 | 444 | 0 | | 141.2 | 39 | 89.3 | 37.1 T |
| 1156 | 5175 | 8.840 | 37 | 57 | 211.2 | 207.2 | 4.05 | 446 | 444 | 2 | | 151.5 | 36 | 90.0 | 38.1 T |
| 1158 | 5084 | 8.840 | 38 | 56 | 211.8 | 206.2 | 5.57 | 445 | 443 | 2 | | 169.1 | 32 | 90.9 | 39.1 T |
| 1160 | 4993 | 8.840 | 39 | 55 | 210.6 | 205.3 | 5.32 | 443 | 443 | -0 | | 167.3 | 33 | 90.8 | 40.1 T |
| 1201 | 4902 | 8.840 | 40 | 54 | 209.3 | 204.3 | 5.00 | 441 | 442 | -2 | | 147.7 | 37 | 89.8 | 41.1 T |
| 1203 | 4811 | 8.840 | 41 | 53 | 206.5 | 203.3 | 3.12 | 443 | 442 | 1 | | 177.5 | 31 | 91.3 | 42.1 T |
| 1205 | 4720 | 8.840 | 42 | 52 | 204.5 | 202.4 | 2.15 | 439 | 441 | -2 | | 167.1 | 33 | 90.8 | 43.1 T |
| 1207 | 4629 | 8.840 | 43 | 51 | 205.9 | 201.4 | 4.50 | 441 | 441 | 0 | | 175.9 | 31 | 91.2 | 44.1 T |
| 1209 | 4538 | 8.840 | 44 | 50 | 202.6 | 200.5 | 2.18 | 441 | 440 | 1 | | 206.1 | 26 | 92.4 | 45.1 T |
| 1211 | 4447 | 8.840 | 45 | 49 | 204.7 | 199.5 | 5.24 | 441 | 440 | 1 | | 169.5 | 32 | 90.9 | 46.1 T |
| 1213 | 4356 | 8.840 | 46 | 48 | 202.5 | 198.5 | 3.97 | 437 | 439 | -2 | | 165.9 | 33 | 90.8 | 47.1 T |
| 1215 | 4265 | 8.840 | 47 | 47 | 202.4 | 197.6 | 4.86 | 439 | 439 | 1 | | 181.6 | 30 | 91.5 | 48.1 T |
| 1217 | 4174 | 8.840 | 48 | 46 | 201.8 | 196.6 | 5.21 | 437 | 438 | -1 | | 223.1 | 24 | 92.9 | 49.1 T |
| 1219 | 4083 | 8.840 | 49 | 45 | 197.3 | 195.7 | 1.65 | 436 | 438 | -2 | | 195.8 | 23 | 92.0 | 50.1 T |
| 1221 | 3992 | 8.840 | 50 | 44 | 197.0 | 194.7 | 2.28 | 439 | 437 | 1 | | 178.0 | 31 | 91.3 | 51.1 T |
| 1223 | 3901 | 8.840 | 51 | 43 | 196.4 | 193.7 | 2.64 | 435 | 437 | -2 | | 202.9 | 27 | 92.3 | 52.1 T |
| 1225 | 3810 | 8.840 | 52 | 42 | 198.1 | 192.8 | 5.35 | 438 | 436 | 2 | | 191.3 | 29 | 91.9 | 53.1 T |

Figure 5-13. Trip Monitor Printout

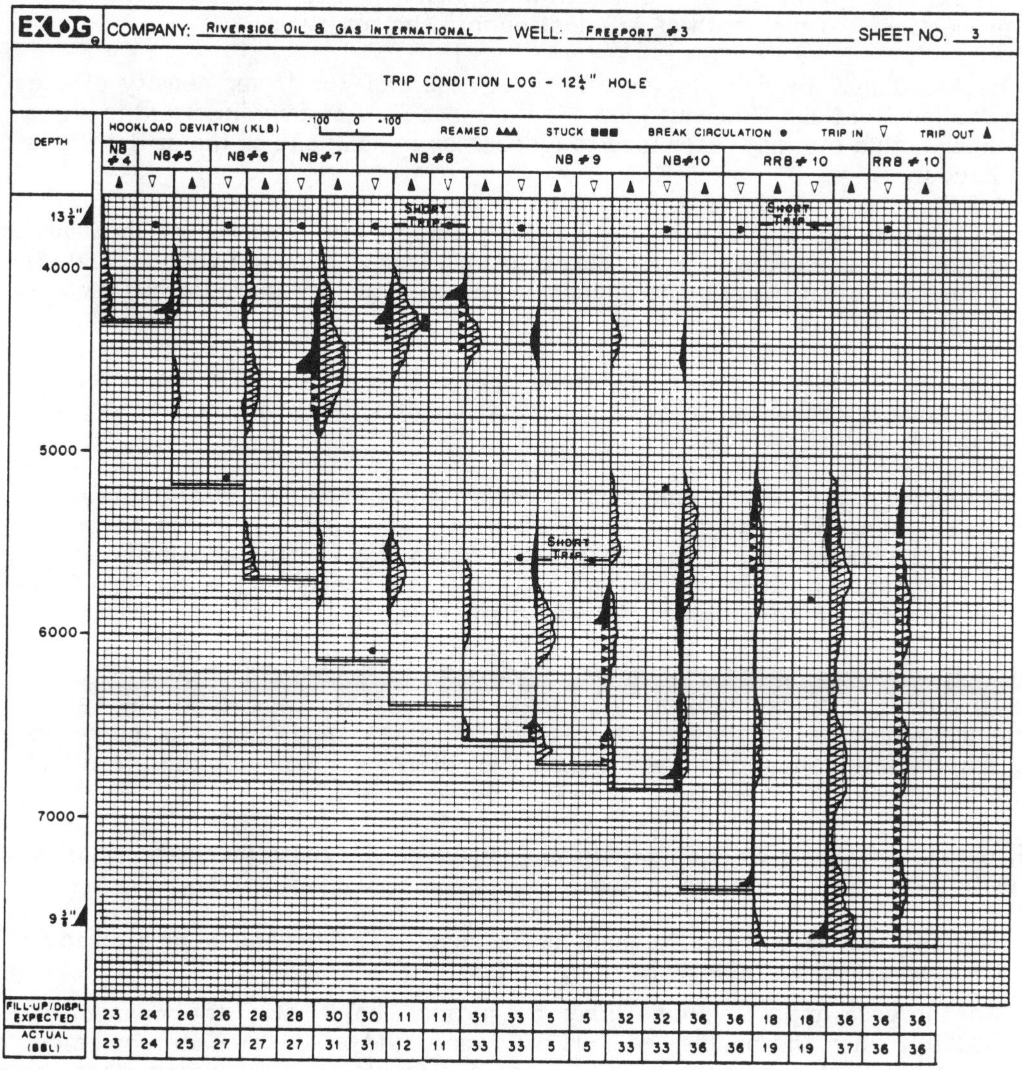


Figure 5-14. Trip Condition Log

fluid is normally lighter than the mud, so continual influx will further lighten the mud column and further reduce bottomhole pressure. This allows the influx rate to increase. Once flow begins, the rate-of-flow increase is proportional to the depth of penetration into the reservoir.

3. Hookload may be seen to increase as a result of the lower density of the invading fluid and fluid-cut mud. If the mass flow of invading fluid is great enough, this too may later lead to a decrease in hookload as the drillstring is lifted by it.

4. An increase in pit volume may be the result of two separate mechanisms: (1) the increased flow rate translates into an increase in mud volume, and (2) if the kick contains gas, the expansion causes a further increase in flowrate and pit volume.

5. A pump pressure decrease occurs as lighter fluid invades the borehole, but becomes noticeable only when the kick material has been displaced some distance up the annulus.

6. A reduction in flowline mud density occurs as the intruding material reaches the surface. This reduction is severe with a gas kick, but it can be either large or unnoticeable with a water kick, depending on the mud density. High gas concentrations are dissolved in kick fluids; thus, as the foreign fluid reaches the surface, high gas shows usually occur.

It is vital that alarms be set on as many of the sensors as possible. However, there are sufficient exceptions to the rule to make it unwise to depend on one factor alone in the sequence of events. For instance, continual mud mixing in the active system may mask a volume increase, and partial returns can so mask the effects of flowrate and volume increase as to make kick detection difficult.

A kick during a connection is signaled by a sequence of events much the same as while drilling:

1. The well may flow when the pump is first shut off. This can be monitored by the flow sensor and the PVT.

2. An increase in pit volume may be noticed only after the connection. Usually, when the pumps are shut off, some mud in the surface equipment flows back into the suction pit. When the levels have stabilized after the pumps are restarted, an increase in level from before the connection indicates that a flow has occurred. This volume should be established at the start of each new job, and reestablished periodically as the well progresses e.g., a 2-bbl increase on a connection may be normal, whereas a 3-bbl rise may be significant.

3. Loss of pump pressure occurs as the annular column becomes lighter due to the invading fluids. This may become noticeable only after successive connections if the flow is slight; however, the flow will increase during each connection.

4. Reduction of mud density at the flowline will eventually occur when the kick fluids reach the surface.

Recognition of kicks during connections requires careful monitoring of the flow sensor in the flowline. After the pumps have been shut down, the flow sensor should indicate an absolute "no-flow" condition. However, on some rigs, a long sloping flowline causes mud to slowly trickle down after the pumps have been shut off. In these cases, an increase in this flow will indicate a kick. Also, a record of flowline mud density will disclose small mud cuts caused during connections, and this may be accompanied by connection gas. Note that connection gas alone is not necessarily an indication of fluid influx during a connection (see paragraph 4.4).

## 5.29 KICK CONTROL

There are three reliable kick control procedures (paragraphs 5.36, .37 and .38). The selection of one to kill a well depends on the amount and type of kick fluids that have entered the well, the rig's equipment capabilities, the minimum fracture pressure in the open hole, and the drilling and operating companies' policies. Determination of the most suitable and safest method (assuming their company policy allows flexibility of procedures determined by the demands of the situation) involves several important considerations, such as

- The time required to execute complex kill prodecures
- The surface pressures that will arise from circulating out the kick fluids
- The downhole stresses that are applied to the formation during the kill process
- The complexity of the procedure itself relative to implementation, rig capabilities and rig crew experience

It is the express responsibility of the Tool Pusher or Operator's Representative to decide which method will be used to kill the well: under no circumstances should the logging geologist become involved in this decision.

Each of these points must be assessed and their relative importance to the kick situation evaluated before implementing the selected method. In the following paragraphs, elaboration of the above points illustrates the reasoning behind their importance on individual situations.

## 5.30 The Time Factor

The total amount of time taken to implement and complete kill procedures is important if the kick is gas because it will percolate up the annulus and increase the annulus pressure, and also because there may be a danger of pipe sticking — especially if a fresh water mud system is in use. Invading saline pore water may cause the mud cake to flocculate, so the bit, stabilizers and collars would be in danger of sticking.

Considerable time is involved in weighting up the mud, but more important is the time necesary for the kill operation to be completed, as the increase in pressures and strains on the well, surface equipment and personnel should be minimized in the interest of safety and cost. Thus, depending on the kick situation, the decision as to what method should be used must be based on these priorities. The kick procedures that involve the least amount of initial waiting time are (1) the two-circulation (or driller's) method and (2) the concurrent method. In both of these procedures, pumping begins immediately after the shut-in pressures are recorded. However, if the time required to weight-up the mud to the desired kill density is less than the time taken for a circulation, then the engineer's (or one-circulation) method would be preferred. In certain situations the extra time increment required for the two-circulation method may be seriously detrimental to hole stability or may cause excessive blowout-preventer wear.

## 5.31 Surface Pressures

If a gas kick is taken, the annular pressures may become alarmingly high during the course of the kill procedure. This is simply due to the properties of a gas bubble as it nears the surface. If expansion is not allowed to occur, severe pressures will be placed on the annulus and surface equipment. For this reason the most reliable well killing procedures utilize a constant drillpipe pressure and variable annular pressure (through a variable choke) method. Figure 5-15 displays the volumetric behavior of gas as it rises up the annulus.

The kill procedure that involves the least surface pressures must be used if the kick tolerance is low. Figure 5-16 shows the different surface pressure requirements for two different kick situations using the one- and two-circulation methods. The first major difference is noted immediately after the drillpipe is displaced with kill mud. When keeping the drillpipe pressure constant with the constant pump rate, the casing pressure begins to decrease as a result of the higher kill mud hydrostatic pressure in the one-circulation procedure. This initial decrease is not seen in the two-circulation procedure: since the mud density has not changed, the casing pressure increases as the gas expansion displaces mud from the hole. The second major surface pressure difference is noted when the gas approaches the surface. The two-circulation method again has the higher pressures as a result of circulation of the original mud density. Also, after one complete circulation has been made, the one-circulation method has killed the well, resulting in zero surface pressure — whereas the two-circulation method still has pressures on the casing that equal the shut-in drillpipe pressure.

## 5.32 Downhole Stresses

This point is the prime concern during kill operations. If the extra stresses imposed by the kick are greater than the minimum fracture pressure in open hole, a fracture will form and an underground blowout may occur. Similarly, a kill procedure which, through its implementation, places high stresses on the wellbore should not be used in preference to another procedure that imposes lower stresses on the wellbore. Reference to the above points and Figure 5-16 illustrate that the one-circulation method places the minimum stresses on the borehole and surface equipment. When a kick is circulated out, the maximum stresses occur very early in the circulation — particularly in deep wells with higher pressures. At any point in the borehole, the maximum stress is imposed when the top of the kick fluid reaches that point.

Generally, in most cases, if fracture and lost circulation do not occur on initial shut-in, they will not occur through the kill process if the correct procedure is chosen and implemented faultlessly. Formulae that enable most of the relevant pressures to be calculated and predicted are presented in paragraph 5.34.

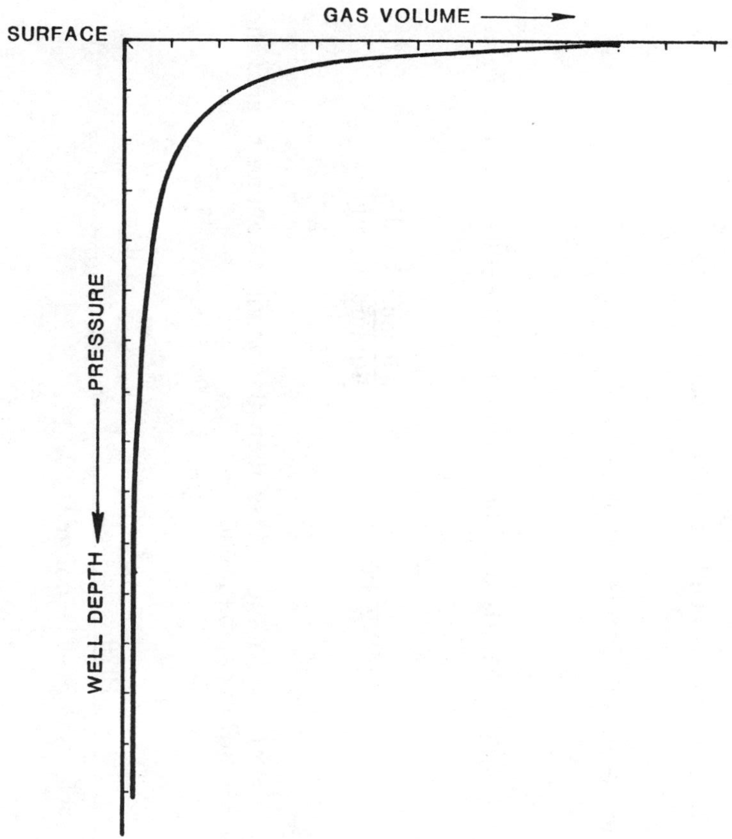


Figure 5-15. Gas expansion on pressure reduction

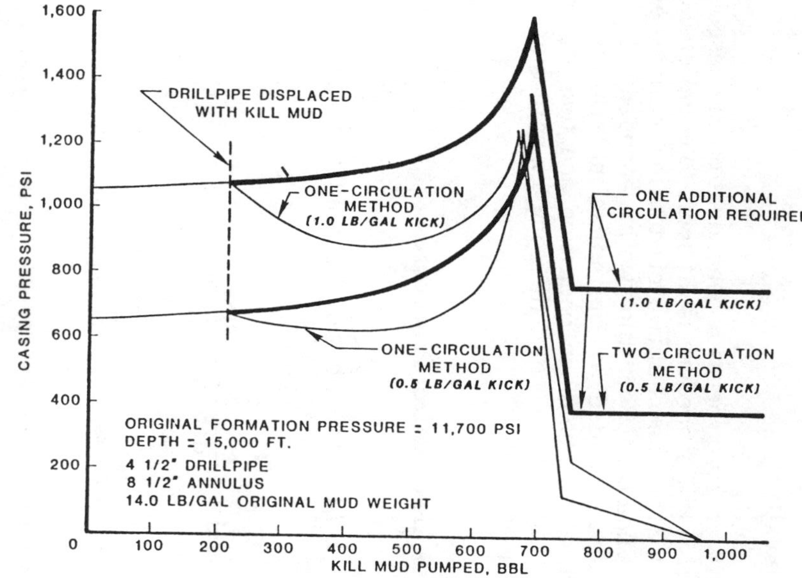


Figure 5-16. Different surface pressures produced during the one- and two-circulation kill methods

### 5.33 Procedural Complexity

The degree of suitability of any process is dependent on the ease with which it may reliably be executed. If a kill procedure is difficult to comprehend and implement, its reliability is negated. The one- and two-circulation methods are simple both in theory and execution. Choice between the two is dependent upon the above points and any other limitations provided by the situation. However, the concurrent method of kill procedure is complex in operation and thus its reliability may be reduced through its intricacy. Because of the complex nature of this method, many operators have discontinued its use.

### 5.34 Formulae Used in Kick Control Procedures

$$\text{Reservoir pressure} = \text{SIDP} + W\ (\text{lb/gal}) * \text{depth (ft)} * 0.0519 \quad \text{psi} \tag{5-31}$$

$$\text{Kill mud density, lb/gal} = \frac{\text{reservoir pressure, psi}}{\text{depth, ft} * 0.0519} \tag{5-32}$$

$$\text{Initial circulating pressure} = \text{System pressure loss at kill rate} + \text{SIDP} \tag{5-33}$$

Normally, pressure losses at various slow pump rates are determined regularly, i.e. every tour change. A usual kill rate would be approximately 30 spm (depending on the pump type). If this information is not available, the pump pressure should be read as soon as the rate has reached the selected kill rate, keeping casing pressure equal to the SICP. The system pressure loss is then found:

$$\text{System pressure loss} = \text{Initial circulating pressure} - \text{SIDP} \tag{5-34}$$

When the kill mud has reached the bit, the final circulating pressure is found from:

$$\text{Final circulating pressure} = \text{system pressure loss} * \frac{\text{kill mud density}}{\text{old mud density}} \tag{5-35}$$

If the kick volume (i.e., pit level gain) is less than the annular volume around the collars, the height of the influx can be found:

$$\text{Length of kick} = \frac{1029}{(\text{hole diameter}^2 - \text{collar O. D.}^2)} * \text{kick volume, bbl} \tag{5-36}$$

If the kick volume exceeds the annular volume around the drill collars, then the influx height can be found:

$$\text{Length of kick} = \text{length of collars} + \frac{1029}{(\text{hole diameter}^2 - \text{pipe O.D.}^2)}$$

$$* \ (\text{kick volume} - \text{annular volume around collars}) \qquad (5\text{-}37)$$

The drillpipe pressure is used as if it represents a downhole pressure gauge. The casing pressure is affected by the type and amount of fluid influx:

$$\text{SICP} - \text{SIDP} = \text{height of influx (ft)} * (\text{mud density, psi/ft} - \text{influx density psi/ft}) \qquad (5\text{-}38)$$

$$\text{SICP} = \text{height of influx (ft)} * (\text{mud density, psi/ft} - \text{influx density, psi/ft}) + \text{SIDP} \qquad (5\text{-}39)$$

$$\text{Density of influx} = \text{mud density, psi/ft} - \frac{\text{SICP} - \text{SIDP}}{\text{height of influx}} \qquad (5\text{-}40)$$

Once the density of the influx fluids is found, the composition may be approximately determined as shown in Figure 5-17.

| Influx Density (psi/ft) | Influx Type |
|---|---|
| 0.05 - 0.2 | gas |
| 0.2 - 0.4 | combination of gas, oil and/or saltwater |
| 0.4 - 0.5 | oil or saltwater |

Figure 5-17. Kick composition from influx gradient

The amount of barite necessary to increase the mud density to the kill density is given by:

$$\text{Sacks of barite per 100 bbl mud} = 1490\ \frac{(\text{kill density - old density})}{35.8 - \text{kill density}} \tag{5-41}$$

and the increase in mud volume caused by weighting up is given by

$$V = \frac{100\,(W_2 - W_1)}{35.8 - W_2} \tag{5-42}$$

where

$V$ = volume increase (bbl)

$W_1$ = initial mud density (lb/gal)

$W_2$ = desired mud density (lb/gal)

It is extremely important to realize that pressures calculated on deviated wells must use vertical depths — not measured depths. Measured lengths, however, must be used for ECD calculations so that the resultant pressure loss is added to the hydrostatic pressure calculated from the vertical depths.

Situations may arise when the casing pressure causes downhole stresses approaching or slightly exceeding the actual or estimated minimum fracture pressure. In this case the well cannot be shut in, and an alternate method of kill control must be attempted. Refer to paragraph 5.39 for an explanation.

The maximum casing pressure at surface is determined by three factors:

- The maximum pressure the wellhead will hold
- The maximum pressure the casing will hold (burst pressure)
- The maximum pressure the formation will hold

Offshore, the limiting factor is usually the latter.

## 5.35 Kick Control Methods

All kick procedures require the knowledge of drillstring geometry, hole geometry, mud density, pump rates and pressure losses, and fracture pressure. Particular information is required prior to initiating kill procedures:

- Circulating pressure at kill rate
- Surface-to-bit time at kill rate, strokes and minutes
- Bit-to-surface time at kill rate, strokes and minutes
- Maximum allowable surface annular pressure
- Formula for calculating kill mud density
- Formula for calculating change in circulating pressure due to the effect of heavier mud
- The Operating Company's policy on safety factors and trip margins

For a well to be killed successfully, the pressure in the formation must be kept under control during the entire operation. Except in cases when the maximum surface annular pressure will be exceeded, this policy should be adhered to. The simplest method of doing this is to control the drillpipe pressure by running the pump at constant speed and controlling the pressure by regulating the choke on the annulus.

Currently, there are three main methods in practice:

- The driller's method (two circulations)
- The wait-and-weight (engineer's) method (one circulation)
- The concurrent method

The ELOS Kick and Kill Analysis (Figure 5-18) and Monitor (Figure 5-19) programs provide calculation of the data required in these procedures and a record of progress during their accomplishment.

**5.36 The Driller's Method (Two Circulations):** When a kick occurs, the drilling crew should proceed as follows:

1. Pick up the kelly and note the position of tool joints in relation to the pipe rams
2. Stop pumps
3. Open the choke line
4. Close the annular preventer or pipe rams
5. Close the choke
6. Record pit gain
7. Record SIDP and SICP when they are stabilized

Calculate the kill-mud density, initial and final circulating pressures, and the kick fluid gradient. If the kick is gas, the bubble may start to percolate up the annulus: this causes a slow rise in pressure on the drillpipe and casing. If the pressures are seen to rise, a small amount of mud is bled from the choke to release the "trapped pressure." This process is repeated until the drillpipe pressure has stabilized.

```
IRWELL VALLEY DRILLING CONSORTIUM : TEST WELL # 1
Kill Analysis Program

Basic Data:
  Depth =                                   11000.0 ft
  Vertical Depth =                          10950.0 ft
  Inclination =                                3.000 degrees
  Fracture Gradient =                          14.54 lb/gal
  At Vertical Depth =                         7950.0 ft
  Mud Density in Hole =                        12.00 lb/gal
  Kick Volume =                                  20 bbl
  Using Pump # 1 at                             5.00 gal/stk
  Slow Circ. Pressure at  30 spm =              500 psi
  Slow Choke Circ. Pressure at  30 spm =        700 psi
  Chokeline Friction =                          200 psi

Volumes:
  Circulating through Choke line instead of Riser/Conductor
   (1) Pipe Capacity =          185 bbl    :    1556 strokes  :   51.9 minutes
   (2) Annulus Volume =        1259 bbl    :   10578 strokes  :  352.6 minutes
   (3) Choke Line Volume =        4 bbl    :      37 strokes  :    1.2 minutes
   (1)+(2)+(3) Volume =        1449 bbl    :   12171 strokes  :  405.7 minutes
   Surface Active Mud Vol. =    250 bbl    :    2100 strokes  :   70.0 minutes
   Total Mud Volume =          1699 bbl    :   14271 strokes  :  475.7 minutes

Kill Data:
  Shut in Drill Pipe Pressure =                 350 psi
  Shut in Casing Pressure =                     400 psi
  Specified Kill Margin =                       100 psi
  Max Permitted Surface Casing Press. =        1049 psi
  Pore Pressure at T.D. =                      7170 psi
  Kill Mud Density Required =                 12.79 lb/gal
  Barite Required = .520 sacks/bbl: Total =     883 sacks
  Mud Volume Increase from Weighting Up =        59 bbl

  Initial Circulating Pressure =                850 psi
  Final Circulating Pressure =                  533 psi
  Required Drill Pipe Pressure Reduction =      317 psi

  Length of Kick =                            239.2 ft
  Vertical Length =                           238.9 ft
  Estimated Kick Fluid Density =               7.97 lb/gal
  Density of Methane at T.D. =                 4.37 lb/gal
  Estimated Gas Percentage in Kick =            22%
```

Plot of Drill Pipe Pressure Against Time and Strokes
When Pumping Kill Mud

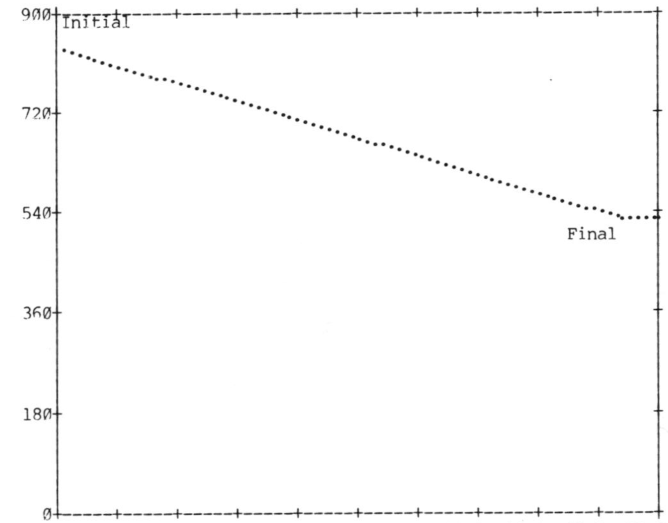


| Time (min) | 0.0 | 5.5 | 11.0 | 16.5 | 22.0 | 27.5 | 33.0 | 38.5 | 44.0 | 49.5 | 55.0 |
|---|---|---|---|---|---|---|---|---|---|---|---|
| Strokes | 0 | 165 | 330 | 495 | 660 | 825 | 990 | 1155 | 1320 | 1485 | 1650 |
| Pressure | 850 | 815 | 783 | 749 | 716 | 682 | 648 | 615 | 581 | 548 | 533 |

Figure 5-18. ELOS Kick and Kill Analysis program

KILL MONITOR

OVERSEAS EXPLORATION COMPANY: TEST WELL #1

UNIT #200    11:47    8/30/80

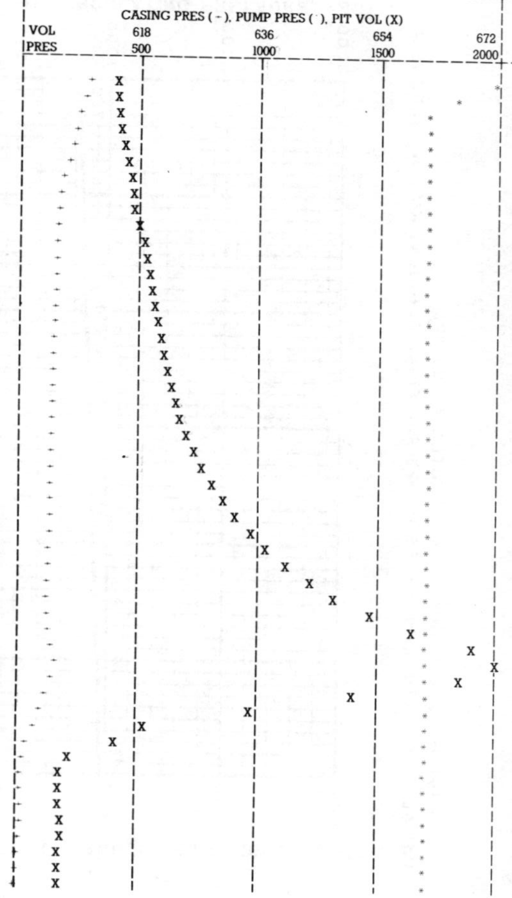


| TIME | CASING PRES | PUMP PRES | PIT VOL | DENSITY IN | DENSITY OUT | FLOW IN | TOTAL BBLS | STROKES | DISPLACEMENT MUD DEPTH | FILE NO. |
|---|---|---|---|---|---|---|---|---|---|---|
| 1150.00 | 269 | 1976 | 615.2 | 11.1 | 9.9 | 673.7 | 16 | 46 | -1129 | 2.1 |
| 1151.00 | 255 | 1804 | 615.7 | 11.1 | 9.8 | 673.7 | 32 | 92 | -2164 | 2.2 |
| 1152.00 | 250 | 1691 | 616.0 | 11.1 | 9.6 | 674.1 | 48 | 138 | 3528 | 2.3 |
| 1153.00 | 237 | 1690 | 616.4 | 11.1 | 9.6 | 673.9 | 64 | 183 | 3288 | 2.4 |
| 1154.00 | 224 | 1690 | 616.9 | 11.1 | 9.7 | 674.5 | 80 | 229 | 3049 | 2.5 |
| 1155.00 | 209 | 1690 | 617.4 | 11.1 | 9.7 | 674.1 | 96 | 275 | 2820 | 2.6 |
| 1156.00 | 182 | 1691 | 617.8 | 11.1 | 9.7 | 674.1 | 112 | 321 | 2682 | 2.7 |
| 1157.00 | 174 | 1689 | 618.2 | 11.1 | 9.7 | 673.8 | 128 | 367 | 2544 | 2.8 |
| 1158.00 | 166 | 1690 | 618.5 | 11.1 | 9.7 | 674.1 | 144 | 413 | 2406 | 3.1 |
| 1159.00 | 157 | 1690 | 618.9 | 11.1 | 9.6 | 674.0 | 160 | 459 | 2268 | 3.2 |
| 1200.00 | 150 | 1690 | 619.3 | 11.1 | 9.5 | 673.8 | 177 | 504 | 2130 | 3.3 |
| 1201.00 | 146 | 1691 | 619.6 | 11.1 | 9.4 | 674.5 | 193 | 550 | 2068 | 3.4 |
| 1202.00 | 143 | 1689 | 620.0 | 11.1 | 9.2 | 674.4 | 209 | 596 | 2008 | 3.5 |
| 1203.00 | 140 | 1690 | 620.4 | 11.1 | 9.1 | 674.2 | 225 | 642 | 1948 | 3.6 |
| 1204.00 | 137 | 1688 | 620.9 | 11.1 | 9.2 | 673.7 | 241 | 688 | 1888 | 3.7 |
| 1205.00 | 135 | 1690 | 621.4 | 11.1 | 9.3 | 673.8 | 257 | 734 | 1830 | 3.8 |
| 1206.00 | 132 | 1691 | 621.9 | 11.1 | 9.2 | 673.8 | 273 | 779 | 1772 | 4.1 |
| 1207.00 | 129 | 1690 | 622.5 | 11.1 | 9.3 | 673.9 | 289 | 825 | 1714 | 4.2 |
| 1208.00 | 127 | 1689 | 623.2 | 11.1 | 9.2 | 674.2 | 305 | 871 | 1656 | 4.3 |
| 1209.00 | 125 | 1688 | 623.9 | 11.1 | 9.2 | 673.7 | 321 | 917 | 1599 | 4.4 |
| 1210.00 | 123 | 1690 | 624.7 | 11.1 | 9.2 | 674.5 | 337 | 963 | 1540 | 4.5 |
| 1211.00 | 121 | 1691 | 625.5 | 11.1 | 9.1 | 674.2 | 353 | 1009 | 1482 | 4.6 |
| 1212.00 | 119 | 1690 | 626.5 | 11.1 | 8.8 | 673.8 | 369 | 1055 | 1424 | 4.7 |
| 1213.00 | 117 | 1692 | 627.6 | 11.1 | 8.3 | 674.5 | 385 | 1100 | 1366 | 4.8 |
| 1214.00 | 116 | 1689 | 628.9 | 11.1 | 8.7 | 674.3 | 401 | 1146 | 1308 | 5.1 |
| 1215.00 | 115 | 1690 | 630.3 | 11.1 | 8.9 | 673.5 | 417 | 1192 | 1250 | 5.2 |
| 1216.00 | 114 | 1688 | 632.0 | 11.1 | 9.0 | 673.9 | 433 | 1238 | 1193 | 5.3 |
| 1217.00 | 114 | 1690 | 633.9 | 11.1 | 9.4 | 673.5 | 449 | 1284 | 1135 | 5.4 |
| 1218.00 | 115 | 1690 | 636.1 | 11.1 | 9.2 | 673.8 | 465 | 1330 | 1077 | 5.5 |
| 1219.00 | 113 | 1691 | 638.8 | 11.1 | 9.2 | 674.0 | 481 | 1376 | 1019 | 5.6 |
| 1220.00 | 109 | 1692 | 642.1 | 11.1 | 9.2 | 674.0 | 497 | 1421 | 961 | 5.7 |
| 1221.00 | 109 | 1690 | 645.6 | 11.1 | 8.9 | 674.1 | 514 | 1467 | 903 | 5.8 |
| 1222.00 | 112 | 1690 | 649.7 | 11.1 | 8.4 | 674.2 | 530 | 1513 | 845 | 6.1 |
| 1223.00 | 116 | 1690 | 654.8 | 11.1 | 7.4 | 673.9 | 546 | 1559 | 787 | 6.2 |
| 1224.00 | 122 | 1690 | 661.4 | 11.1 | 7.3 | 674.3 | 562 | 1605 | 729 | 6.3 |
| 1225.00 | 132 | 1688 | 670.2 | 11.1 | 6.3 | 673.8 | 578 | 1651 | 671 | 6.4 |
| 1226.00 | 147 | 1689 | 672.0 | 11.1 | 5.2 | 674.5 | 594 | 1697 | 613 | 6.5 |
| 1227.00 | 121 | 1690 | 667.7 | 11.1 | 5.4 | 674.5 | 610 | 1742 | 555 | 6.6 |
| 1228.00 | 94 | 1691 | 651.7 | 11.1 | 5.2 | 674.1 | 626 | 1788 | 502 | 6.7 |
| 1229.00 | 67 | 1690 | 635.6 | 11.1 | 5.8 | 673.5 | 642 | 1834 | 455 | 6.8 |
| 1230.00 | 40 | 1690 | 619.6 | 11.1 | 7.2 | 674.0 | 658 | 1880 | 407 | 7.1 |
| 1231.00 | 18 | 1692 | 615.1 | 11.1 | 7.8 | 673.6 | 674 | 1926 | 359 | 7.2 |
| 1232.00 | 15 | 1688 | 608.1 | 11.1 | 8.1 | 674.1 | 690 | 1972 | 312 | 7.3 |
| 1233.00 | 12 | 1690 | 607.2 | 11.1 | 8.2 | 674.0 | 706 | 2017 | 264 | 7.4 |
| 1234.00 | 10 | 1689 | 607.0 | 11.1 | 8.3 | 674.5 | 722 | 2063 | 216 | 7.5 |
| 1235.00 | 9 | 1691 | 607.0 | 11.1 | 8.8 | 674.0 | 738 | 2109 | 169 | 7.6 |
| 1236.00 | 8 | 1690 | 607.1 | 11.1 | 8.8 | 673.5 | 754 | 2155 | 121 | 7.7 |
| 1237.00 | 7 | 1692 | 607.0 | 11.1 | 10.0 | 673.6 | 770 | 2201 | 73 | 7.8 |
| 1238.00 | 6 | 1688 | 607.0 | 11.1 | 9.9 | 674.3 | 786 | 2247 | 26 | 8.1 |
| 1239.00 | 5 | 1690 | 607.0 | 11.1 | 9.9 | 674.1 | 802 | 2293 | 0 | 8.2 |
| 1240.00 | 0 | 1690 | 607.1 | 11.1 | 10.8 | 673.9 | 818 | 2338 | 0 | 8.3 |

Figure 5-19. ELOS Kick and Kill Monitor program

The first circulation is performed using the original mud. The choke is opened slightly at the same time the pumps are started up to the kill rate. When the pumps have reached the kill rate, the choke is manipulated to keep the pressure on the drillpipe at the original SIDPP plus the circulation pressure. As the kick fluids approach the surface, the annular pressure will rise drastically if the kick is gas. If the kick is saltwater, the annular pressures will fall slightly.

When all the influx has been circulated out, the pump is stopped and the choke closed. The drillpipe pressure should be the same as the casing pressure.

During the first circulation the mud density in the pits should have been increased to the necessary kill density, and the kill mud is circulated during the second circulation. The choke is opened slowly and the pump speed is increased to the kill rate, as the annular pressure is kept constant. The annular pressure is kept constant by manipulating the choke until the kill mud has reached the bit. The drillpipe pressure will decrease during this operation from the initial circulating pressure to the final circulating pressure (Figure 5-20). It is good practice at this point to close the well in. The drillpipe pressure should fall to zero; if it doesn't, a few more barrels should be pumped to be sure the kill mud has reached the bit. If the drillpipe pressure is still greater than zero when the pump is stopped and the choke is closed, the kill control figures should be checked. Pumping is restarted, but now the drillpipe pressure is kept constant as the kill mud displaces that in the annulus. When the kick fluids and original mud have been displaced the choke will be wide open, the pump should be shut down, and the well observed for flow. The well will be dead, the mud should be circulated to condition the hole, and at the same time the trip margin should be added. Figures 5-21 and 5-22 show the pressure behaviors as the first and second circulations are performed.

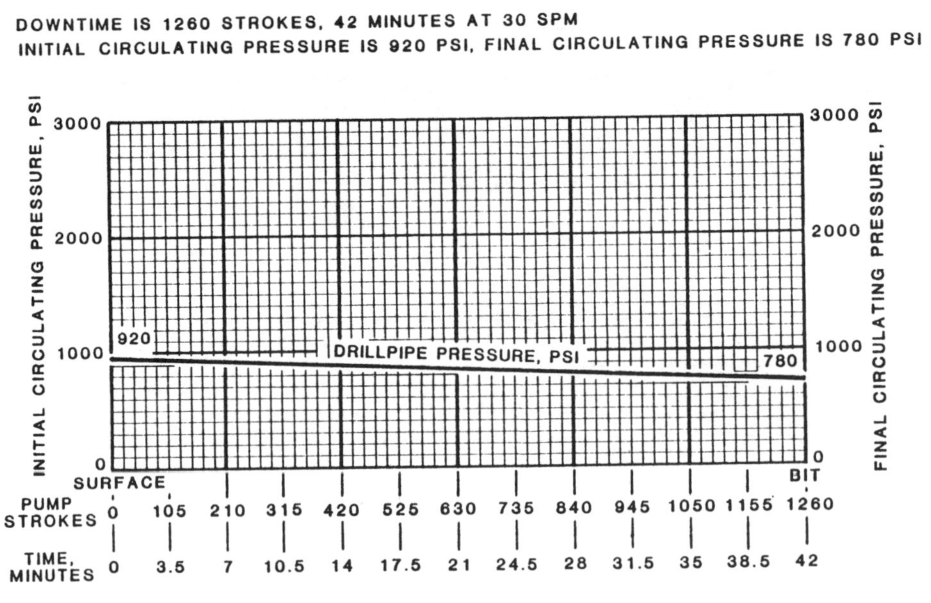


Figure 5-20. Drillpipe/pressure plot when kill mud is pumped down the drillpipe

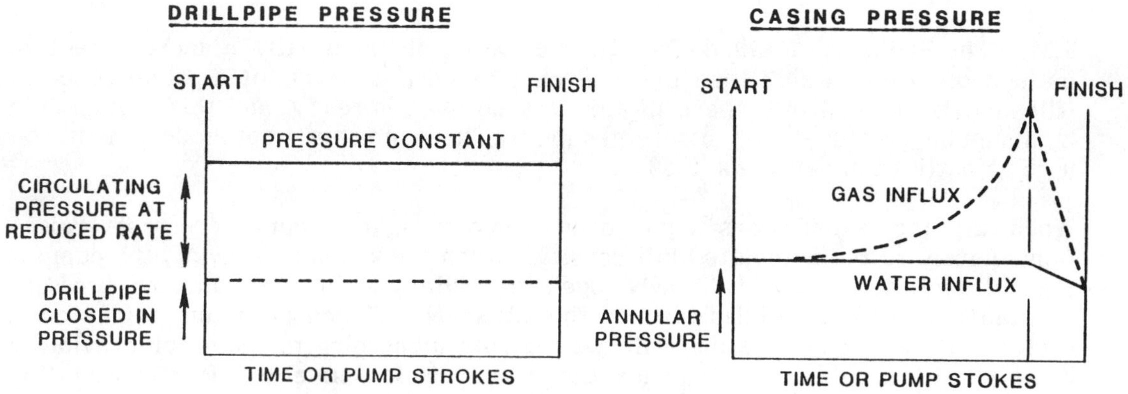


Figure 5-21. First circulation pressures during the driller's method. Drillpipe pressure is kept constant by gradually opening the choke. As gas reaches the surface, it may be necessary to slow the pump rate in order to keep the drillpipe pressure constant

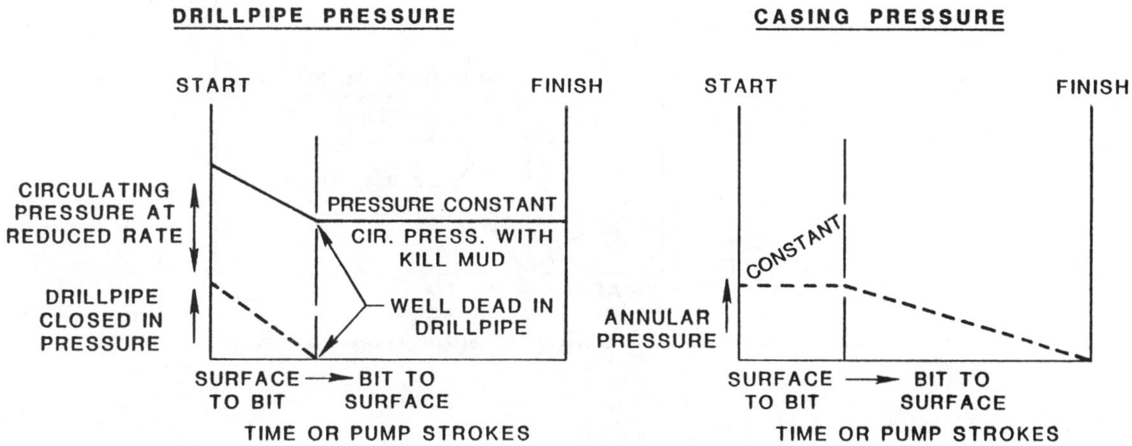


Figure 5-22. Second circulation during driller's method: kill mud is pumped around

**5.37 The Engineer's Method (One Circulation):** This is usually a more effective method of killing a kick than the driller's method if time is not a prime concern. Kill mud is pumped into the drillpipe as soon as it is ready, and this reduces the high annular pressures that signify gas kicks. The same shut-in procedure should be used as outlined in paragraph 5.36.

When all the calculations have been performed, the mud density is raised immediately to the calculated kill density. When the kill mud is ready, the pump is started and the choke is slowly opened, while keeping the annular pressure constant, until the pump has reached the kill speed. The choke is then regulated in such a way as to decrease the drillpipe pressure (according to the graph previously drawn) until the kill mud reaches the bit, at which time the final circulating pressure is reached.

Pumping is continued, holding the drillpipe pressure constant by adjusting the choke. When the kick fluids have been displaced, and further volume has been displaced equal to the pipe volume, the well should be dead and checked for flow. Further circulations should then be performed to condition the hole and to add the trip margin. Figures 5-23 and 5-24 show the variation of drillpipe and casing pressures as the kill procedure is implemented. Figure 5-25 shows diagrammatically the displacement of the original mud with kill mud, with example pressures.

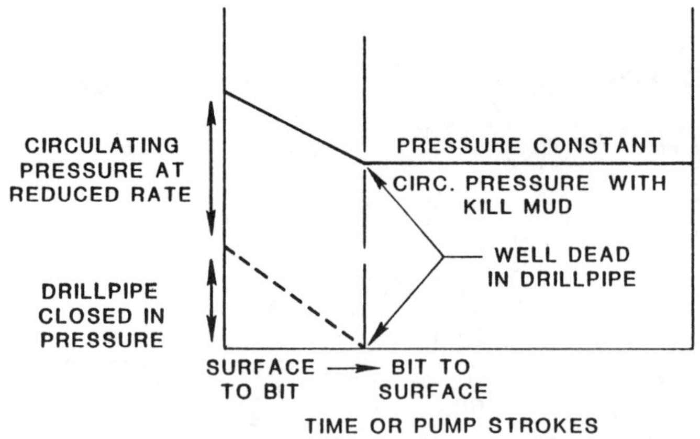


Figure 5-23. Drillpipe/pressure curve during engineer's kill method

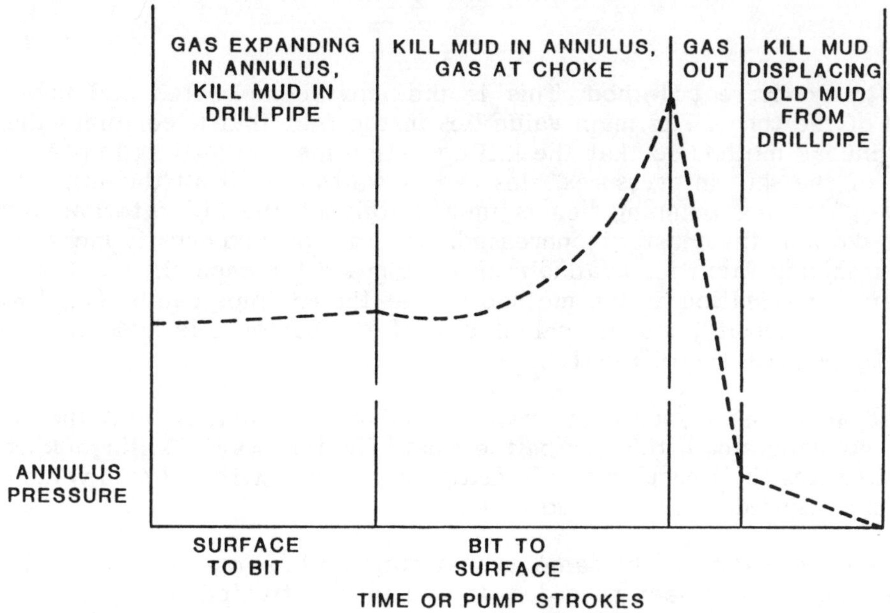


Figure 5-24. Annulus pressure curve during engineer's kill method. Note casing pressure reduction after kill mud has reached the bit

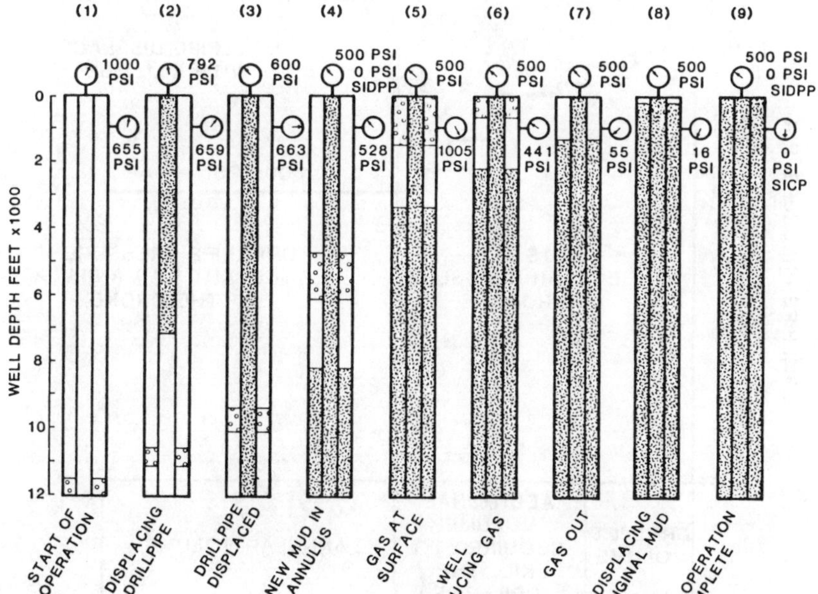


Figure 5-25. Schematic of example pressures and fluid positions during the engineer's kill method

**5.38 The Concurrent Method:** This is the most complicated and unpredictable method of the three. Its main value lies in the fact that it combines the driller's and engineer's method so that the kill operation may be initiated upon immediate receipt of the shut-in pressures. Instead of waiting until all the surface mud has been weighted up, pumping begins immediately at the kill rate and the mud is pumped down as the density is increased. The rate of mud density increase depends upon the mixing facilities available on the rig and the capability of the personnel. The main complication of the method is that the drillpipe can be filled with muds of increasing density, making calculation of the bottomhole hydrostatic pressure (and drillpipe pressure) difficult.

Provided there is adequate supervision and communication and the method is completely understood, this can be the most effective way of killing a kick. Figure 5-23 illustrates the irregularities in drillpipe pressure with kill mud volume, caused by the increasing density of the kill mud.

The shut-in procedure is the same as that outlined in paragraph 5.36. When all the kick information has been recorded, the pump is activated slowly until the initial circulating pressure has been reached at the designated kill rate. The mud should be weighted up at maximum possible rate, and, as the mud density changes in the suction tank, the choke operator is informed. The pump strokes already passed are checked on the drillpipe pressure chart when the new density is pumped, adjusting the choke to suit the new drillpipe conditions as prerecorded on the surface-to-bit graph (Figures 5-20 and 5-26). When the final kill mud reaches the bit, the final circulating pressure will be reached — and from this point on, the drillpipe pressure should be kept constant until the operation is completed.

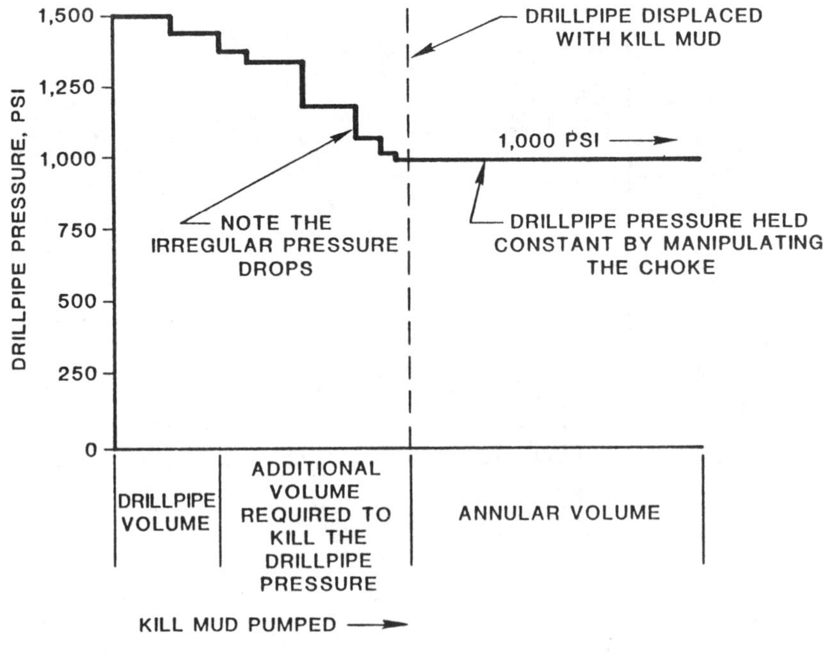


Figure 5-26. Typical irregular drillpipe pressure reductions during concurrent method

## 5.39 KICK TOLERANCE

Kick tolerance is defined as the maximum Formation Balance Gradient that may be encountered and a kick taken at the present depth and with the present mud density and the well shut-in without downhole fracturing resulting. If the pore, fracture (actual or theoretical) and mud hydrostatic pressures are continually monitored, then in the majority of cases kick tolerance may be closely estimated. The limit of this pressure is usually set by the minimum fracture pressure in open hole.

It is of paramount importance that the estimated kick tolerance is not exceeded. A well cannot be drilled safely if kick tolerance is exceeded because, if the kick is taken, there is considerable chance that an underground blowout will occur if the well is shut in. The maximum surface pressure possible will be a function of the mud density and the depth and Fracture Pressure Gradient of the weakest formation in open hole.

$$SICP_{max} = 0.0519 * (FG_{min} - W) * D_f \qquad (5\text{-}43)$$

where

$SICP_{max}$ = maximum shut-in casing pressure (psi)

$FG_{min}$ = fracture pressure gradient of weakest formation (lb/gal)

$W$ = mud density (lb/gal)

$D_f$ = vertical depth of weakest formation (ft)

The Formation Balance Gradient which would produce this maximum shut-in casing pressure may be calculated for a particular depth and mud density.

$$.0519*K*D_B = SICP_{max} + .0519 * W * D_B \qquad (5\text{-}44)$$

Which can be restated as:

$$K = \left[\frac{D_F}{D_B}(FG_{min} - W) + W\right] \qquad (5\text{-}45)$$

where

$K$ = kick tolerance (lb/gal)

$D_B$ = vertical depth of bit (ft)

This defines the maximum formation balance gradient that may be encountered at that depth so that the well may be shut in without exceeding the lowest fracture gradient. However, this expression assumes that the kick will be detected and the well shut-in with zero influx of formation fluid. In reality, the kick tolerance will be reduced by a term involving the density and volume of invading fluid. This may be calculated from:

$$K_{min} = \left[\frac{D_F}{D_B}(FG_{min} - W)\right] - \left[\frac{L_k}{D_B}(W - W_k)\right] + W \qquad (5\text{-}46)$$

where

$K_{min}$ = minimum kick tolerance (lb/gal)

$L_k$ = length of kick (ft)

$W_k$ = density of kick fluids (lb/gal)

Kick fluid density will vary with the nature of the invading fluid but will be a minimum in a gas kick. The expected influx prior to detecting a kick will depend upon the resulution of the pit-level monitoring apparatus, that is the equipment design and the height-to-surface area ratio of the pit. It will also depend upon the speed and efficiency of the rig crew and equipment. This cannot be calculated but must be determined by tests, that is pit drills.

A large influx of gas significantly decreases the kick tolerance. Hence if the kick tolerance with no influx is being approached, the minimum kick tolerance may be determined by the resolution of the pit-level monitoring apparatus. For safety's sake it should be assumed that a kick will be gas. Thus the minimum noticeable pit gain (i.e. 15 bbl), should be added to the estimated pit gain that will occur (due to the time lapse from first observing the flow) to when the well is finally shut in. This shut-in period is critical, and usual practice is to run through a pit drill and hang-off procedure regularly. A minimum time of, say, 1 minute may allow a further 20-bbl pit rise. Thus (in this case) the total minimum pit gain before the well could be shut in would be 35 bbl. This value defines the minimum expected kick length in a particular hole section; thus, with a particular mud density, the reduction of kick tolerance due to gas influx may also be continually estimated. For example:

If 12-1/4-inch hole is being drilled and 642 feet of 8-inch collars are being used,

$$\text{length of kick} = \frac{1029}{(12\ 1/4)^2 - 8^2} * 35 \qquad (5\text{-}36)$$

$$= 418 \text{ ft}$$

With 15 lb/gal mud in the hole, and assuming a kick would be gas having a density of 2 lb/gal, at a current depth of 15,000 ft and a minimum fracture pressure gradient of 16.0 lb/gal at 7500 ft,

$$K_{min} = \left[\frac{D_F}{D_B}(FG_{min} - W)\right] - \left[\frac{418}{15,000}\right](15.0 - 2) + W$$

$$= \left[\frac{2500}{15000}(16.0 - 15.0)\right] - 0.36 + 15.0 \text{ lb/gal}$$

$$K_{min} = 15.14 \text{ lb/gal}$$

If kick tolerance calculations did not take into account the minimum expected influx,

$$K = \frac{7500}{15,000}(16.0 - 15) + 15.0$$

$$K = 15.50 \text{ lb/gal}$$

which may appear to be a reasonable safety margin if the pore pressure at 15,000 ft was estimated to be just overbalanced. But if the minimum kick influx is taken into account, the actual kick tolerance would be only 15.14 lb/gal. Further, if for some reason a kick was taken and a total pit gain of 50 bbl occurred, so that the kick length was 598 ft, then the total kick tolerance would be

$$K_{min} = 0.5 - \frac{598}{15,000} * (15.0 - 2.0) + 15.0$$

$$= 0.5 - 0.52 + 15.0$$

$$K_{min} = 14.98 \text{ lb/gal}$$

This example serves to illustrate the highly important function of correcting the kick tolerance for influx. In this case, kick tolerance is less than mud density (i.e. the well cannot be shut in) as the shut-in casing pressure would be such that the formation would be fractured at 7500 ft and an underground blowout would occur.

<u>It is vitally important that the 'minimum expected influx' is utilized in kick tolerance estimations in order that the well may be safely shut in if a gas kick is taken.</u>

## IMPORTANT NOTE

A Minimum Kick Tolerance ($K_{min}$, in Equation 5-46), corrected for invading fluid volume and density, may be calculated and reported only when the volume and density are specified by the drilling supervisor, oil company standard operating procedure, or a regulatory agency. In such circumstances the GEMDAS Operator or P.E.G. may assist the drilling supervisor in determining the volume and density estimates to be used, but the authorization for their use must come from a representative of the oil company.

The "minimum expected influx" will vary during the progress and must be re-established with regular pit drills. When it is reported, for example on the daily GEMDAS Logging Report Form, the expected influx volume and density must be reported with it, for example:

Kick Tolerance: <u>15.2 lb/gal w/25 bbl/2 lb/gal influx</u>

Plots of Kick Tolerance, for example on the Pressure Evaluation Log, should be of the true, zero influx, Kick Tolerance (K) only.

Operational situations may arise which cause the kick tolerance to be exceeded. If a pit gain occurs such that it will bring the actual kick tolerance to 0.1 lb/gal above current mud density, the well should not be killed by the accepted methods if the kick is calculated to be gas. After the shut-in readings have been taken, it may be possible to kill the well by slowly pumping a large barite or gunk plug down the well, or by pumping the original mud at a high rate against a small choke backpressure.

Gas does not start to expand significantly until the pressure on the influx is reduced to less than 6000 psi. Thus, if a gas kick is taken, gas expansion and pressure increase will not be rapid until the influx has been circulated (or percolated) up to a level in the borehole where the pressure is less than 6000 psi. For example, if kick tolerance is becoming marginal in a deep well where casing is set at 7500 ft, as long as the mudweight is greater than 15.4 lb/gal, gas expansion will not occur until the gas is inside the casing.

The accuracy of the calculation of kick tolerance can be seen to be dependent upon the accuracy of the majority of the geopressure evaluation techniques. In actuality, kick tolerance is the goal toward which the geopressure evaluation service is directed. Consider the terms that make up the relationship, and their description:

$D_F$ = Vertical depth of the weakest formation; necessitates knowledge of formation type (i.e., Poisson's ratio) and pore pressure.

$D_B$ = Vertical depth of the bit; if the hole is deviated, we need to be able to calculate (through survey analysis) the vertical depth.

$FG_{min}$ = Minimum fracture pressure gradient necessitates estimation of the overburden pressure gradient, pore pressure gradient (for $\sigma_1'$) interpretation of the first fracture test in compact formation and back-calculation of $\sigma_t$. Then, it requires monitoring of pore pressure changes, lithological changes, and overburden pressure extrapolation as the well progresses in order to delineate the weakest formation in the borehole, and to estimate its fracture pressure.

The Exlog P.E.G.s and GEMDAS operators must be aware, either through their own experiences or — in the case of the inexperienced geologist — through this manual, of the importance of their measurements and interpretations. Communication of these results to the client must be concise and unambiguous in order that full use may be made of them. With current safety levels coming under scrutiny, government agencies are involving themselves in rig practices that have the potential to endanger lives and the environment. The establishment of kick tolerance safety levels is one of these criteria, and in some countries these laws have been established for some years. As more countries follow suit, under the impetus of more energetic exploration in hazardous areas, kick tolerance calculations will become perhaps one of the most important aspects of Exlog's P.E.G. and GEMDAS involvement at the wellsite.

## 5.40 "DIFFERENTIAL" KICK TOLERANCE

It is conventional in the oilfield to compute and report pressure - related quantities as gradients relative to the flowline (see Section 3). This is a convenience allowing direct comparison of the quantities to the mud density currently in use.

In some areas of the world this convention is discarded when reporting kick tolerance. A figure is reported which we will call "differential" kick tolerance which is the actual kick tolerance minus the actual mud density.

$$\text{kick tolerance, } K = \left[\frac{D_F}{D_B}(FG_{min} - W)\right] + W \qquad (5\text{-}45)$$

"differential" kick tolerance = (K - W)

$$\Delta K = \frac{D_F}{D_B}(FG_{min} - W) \qquad (5\text{-}47)$$

minimum kick tolerance

$$K_{min} = \left[\frac{D_F}{D_B}(FG_{min} - W)\right] - \left[\frac{L_K}{D_B}(W - W_K)\right] + W \qquad (5\text{-}46)$$

"differential" minimum kick tolerance = $(K_{min} - W)$

$$\Delta K_{min} = \left[\frac{D_F}{D_B}(FG_{min} - W)\right] - \left[\frac{L_K}{D_B}(W - W_K)\right] \qquad (5\text{-}48)$$

The rationale for this method of reporting is that this quantity will decline as the hole is deepened and when mud density is increased, reaching zero at the point at which the well can no longer be shut in. This is a dramatic representation of declining, safety margins. On the other hand, it removes the direct comparability of the quantity, especially when plotted.

Reporting of kick tolerance in this form should not be encouraged. P.E.G.s and GEMDAS operators who are requested to do so by a client must of course comply but should be careful to discriminate between the differential being reported and the actual term being plotted on the Pressure Evaluation Log.

## 5.41 REFERENCES


Anderson, E. M., 1942, The Dynamics of Faulting, Oliver and Boyd, London.

Anderson, R. A., D. S. Ingram and A. M. Zanier, 1973, Determining Fracture Pressure Gradients from Well Logs, J.P.T., Nov.

Bellotti, P., and D. Giacca, 1978, Pressure Evaluation Improves Drilling Programs, O & GJ, Sept 11.

Bradley, W. B., 1979(a), Mathematical Concept -- Stress Cloud -- Can Predict Borehole Failure, O&G J., Feb.

Bradley, W. B., 1979 (b), Predicting Borehole Failure Near Salt Domes, O&G J., Apr.

Chenevert, M. E., and L. J. McClure, 1978, How to Run Casing and Open Hole Pressure Tests, O&G J., Mar.

Christman, S. A., 1973, Offshore Fracture Gradients, J.P.T., SPE Paper 4133.

Costley, R. D., 1967, Hazards and Costs Cut by Planned Drilling Programs, World Oil, Oct.

Eaton, B. A., 1969, Fracture Gradient Prediction and Its Application in Oilfield Operations, J.P.T., Oct.

Fairhurst, C., 1960, On the Determination of the State of Stress in Rock Masses, SPE Paper 1062.

Fowler, P.T., 1980, Telling Live Basins from Dead Ones By Temperature, World Oil, May.

Goins, W. D. Jr., 1968, Guidelines for Blowout Prevention, World Oil, Oct.

Hafner, W., 1951, Stress Distributions and Faulting, Bull. Geol. Soc. Am., v. 62.

Haimson, B., and C. Fairhurst, 1969, Hydraulic Fracturing in Porous-Permeable Materials, J.P.T., SPE Paper 2354.

Haimson, B., and J. N. Edl, 1972, Hydraulic Fracturing of Deep Wells, SPE Paper 4061.

Handin, J. et al, 1963, Experimental Deformation of Sedimentary Rocks Under Confining Pressure: Pore Pressure Tests, AAPG Bull., v. 47, n. 5.

Hubbert, M. K., 1945, Strength of the Earth, AAPG Bull., v. 29, n. 11.

Hubbert, M. K., and D. G. Willis, 1957, Mechanics of Hydraulic Fracturing, Trans. AIME, v. 210.

Jaeger, J. C., and N.G.W. Cook, 1976, Fundamentals of Rock Mechanics, Chapman and Hall, 2nd ed., London.

Jeffreys, H., 1952, The Earth, Cambridge, 3rd ed.

Kendall, H. A., 1977, How to Control Deep Critical Wells, Pet. Eng., Mar.

MacPherson, L. A., and L. N. Berry, 1972, Prediction of Fracture Gradients from Log Derived Moduli, The Log Analyst, Sep-Oct.

Matthews, W. R., and J. Kelly, 1967, How to Predict Formation Pressure and Fracture Gradient, O&G J., Feb. 20

Pennebaker, E. S., 1968, An Engineering Interpretation of Seismic Data, SPE Paper 2165, presented at the 43rd Ann Fall Meeting of SPE-AIME, Houston.

Pilkington, P. E., and H. A. Neihaus, 1975, Exploding the Myths About Kick Tolerance, World Oil, Jun.

Pilkington, P. E., 1978, Fracture Gradient Estimates in Tertiary Basins, Pet. Eng., May.

Rehm, B., 1976, Deep Water Drilling Poses Special Pressure Control Problems, O&G J, May 3.

Reynolds, E. B., 1970, Predicting Overpressured Zones with Seismic Data, World Oil, v. 171, Oct.

Taylor, D. B., and T. K. Smith, 1970, New Fracture Gradients Help Cut Costs Offshore, World Oil, Jun.

U. S. Bureau of Reclamation, 1953, Physical Properties of Some Typical Foundation Rocks, Concrete Laboratory Rpt. SP-39.

Weurker, R.G., 1963, Annotated Tables of Strength and Elastic Properties of Rocks, SPE Reprint Series, n.6.

# APPENDIX A
# FORMULAE

| Formulae and Paragraph | Equation Number |
|---|---|

**Formation Volume Factor, 2.3**

$$B = (1 - dV_{wp}) * (1 + dV_{wt}) \qquad (2\text{-}1)$$

**Darcy's Law, 2.10**

$$Q = K \frac{A}{\mu L} * \Delta P, \text{ cm}^3/\text{sec} \qquad (2\text{-}2)$$

**Hydrostatic Pressure, 3.3**

$$P = 0.0519 * W * D, \text{ psi} \qquad (3\text{-}1)$$

$$P = 0.0098 * W * D, \text{ kPa} \qquad (3\text{-}6)$$

**Bulk Density, 3.4**

$$\rho_b = \emptyset \rho_F + (1-\emptyset)\, \rho_m, \text{ g/cc} \qquad (3\text{-}8)$$

$$\emptyset = \frac{\rho_m - \rho_b}{\rho_m - \rho_f} \qquad (3\text{-}9)$$

**Overburden Gradient, 3.4**

$$S = 0.433 * \rho_b \text{ average} * D, \text{ psi} \qquad (3\text{-}10)$$

$$OBG = \frac{\sum 0.433 * \rho_b \text{ average} * D}{\sum D}, \text{ psi/ft} \qquad (3\text{-}11)$$

Effective Overburden Pressure, 3.6

$$\sigma_1' = S - P, \text{ psi} \tag{3-12}$$

Annular Pressure Loss, 3.7

$$\delta p = \frac{L * YP}{A * (I.D. - O.D.)} + \frac{PV * L * V}{B * (I.D. - O.D.)^2}, \text{ psi} \tag{3-14}$$

$$v = \frac{24.51 * \text{gallons per minute}}{(I.D.^2 - O.D.^2)}, \text{ ft/min} \tag{3-15}$$

$$ECD = W + \frac{\sum \delta p}{0.0519 * D}, \text{ lb/gal} \tag{3-16}$$

$$BHCP = ECD * D * 0.0519, \text{ psi} \tag{3-17}$$

$$\delta p = \frac{L \tau}{300 \, (I.D. - O.D.)}, \text{ psi} \tag{3-18}$$

$$W_{trip} \leq W - \frac{\sum \delta P}{0.0519 * D}, \text{ lb/gal} \tag{3-19}$$

Well Control, 3.8

$$P = P_{md} + SIDP = P_{ma} + P_k + SICP, \text{ psi} \tag{3-21}$$

$$P = (0.0519 * D * W) + SIDP, \text{ psi} \tag{3-22}$$

$$W' = W + \frac{SIDP}{0.0519 * D}, \text{ psi} \qquad (3\text{-}23)$$

Interval Transit Time, 4.2

$$T = \frac{1}{V} * 1000, \text{ microsec/ft} \qquad (4\text{-}1)$$

Differential pressure, 4.4

$$\Delta P = (W * D * 0.0519) - (FBG * D * 0.0519), \text{ psi} \qquad (4\text{-}2)$$

Gas-Cut Mud, 4.5

$$G_v = \left(\frac{d}{24}\right)^2 * \frac{\pi * R}{60} * \emptyset * Sg * 7.48, \text{ gal/min} \qquad (4\text{-}3)$$

$$G_{va} = G_v * \frac{P}{14.7}, \text{ gal/min} \qquad (4\text{-}4)$$

$$W_1 = \frac{\text{Mud (gpm)}}{\text{Mud (gpm)} + \text{Gas (gpm)}} * W_2, \text{ lb/gal} \qquad (4\text{-}5)$$

$$\Delta P = 14.7 \left(\frac{W_2 - W_1}{W_1}\right) \ln \left(\frac{3.53 * W_2 * D}{1000}\right), \text{ psi} \qquad (4\text{-}6)$$

D-exponent, 4.9

$$d = \frac{\log \left(\frac{R}{60N}\right)}{\log \left(\frac{12W}{10^6 B}\right)} \qquad (4\text{-}7)$$

$$dxc = d * \frac{N.FBG}{ECD} \tag{4-8}$$

$$dxc = \frac{\log\left(\frac{R}{18.29N}\right)}{\log\left(\frac{W}{14.88B}\right)} * \frac{N.FBG}{ECD}, \text{ metric} \tag{4-10}$$

$$P_o = P_n * \frac{Dxc_n}{Dxc_o}, \text{ lb/gal} \tag{4-11}$$

Bulk Density, 4.11

$$\text{Bulk Density} = \frac{8.34}{16.68 - W_2}, \text{ g/cc} \tag{4-13}$$

Shale Factor (C.E.C.), 4.12

$$\text{Shale Factor} = \frac{100}{\text{Sample Mass, g}} * \text{vol, ml} * \text{(normality of methylene blue solution)} \tag{4-14}$$

$$\text{True Shale Factor} = \frac{100}{100 - \text{Carbonate, \%}} * \text{(apparent shale factor)} \tag{4-15}$$

Temperature, 4.16

$$G = 100 \frac{\left(T_{F_2} - T_{F_1}\right)}{\left(D_2 - D_1\right)}, \text{ }^{o}\text{C/100 Ft} \tag{4-16}$$

$$GF = \frac{G}{G_n} \tag{4-17}$$

$$T = T_F - C \log \frac{(t_c + t_L)}{t_L}, \text{ }^{o}\text{C} \tag{4-18}$$

$$\frac{T_f(t_2-t_1) + [(T_1*t_1) - (T_2*t_2)]}{(T_2 - T_1)} = \frac{T_f(t_3-t_1) + [(T_1*t_1) - (T_3*t_3)]}{(T_3 - T_1)} \quad (4\text{-}20)$$

Mud Resistivity, 4.14

$$R = \frac{1000}{C}, \text{ ohmmeters} \quad (4\text{-}21)$$

Sonic Log, 4.16

$$T_c = \sqrt{\frac{\rho(1+\mu)}{3M_b(1-\mu)}} \quad (4\text{-}22)$$

$$\emptyset = \frac{\Delta t - \Delta t_m}{\Delta t_f - \Delta t_m} \quad (4\text{-}23)$$

Equivalent Depth Method, 4.16

$$P = OBG_a * D_a - D_n\ (OBG_n - N.FBG), \text{ lb/gal} \quad (4\text{-}24)$$

Hubbert & Willis, 5.7

$$F = \frac{(S-P)}{3} + P, \text{ psi} \quad (5\text{-}1)$$

$$\frac{F}{D} \approx \frac{\frac{S}{D} + \frac{2P}{D}}{3}, \text{ psi/ft} \quad (5\text{-}2)$$

Matthews & Kelly, 5.7

$$\frac{F}{D} = \frac{P}{D} + ki\,\frac{\sigma}{D}, \text{ psi/ft} \quad (5\text{-}3)$$

Eaton, 5.7

$$\frac{F}{D} = \left(\frac{S}{D} - \frac{P}{D}\right)\left(\frac{\mu}{1-\mu}\right) + \frac{P}{D}\text{, psi/ft} \qquad (5\text{-}5)$$

Anderson et al, 5.7

$$\frac{F}{D} = \left(\frac{2\mu}{1-\mu}\right) * \frac{S}{D} + \left(\frac{1-3\mu}{1-\mu}\right) * \frac{\alpha P}{D}\text{, psi/ft} \qquad (5\text{-}7)$$

$$\alpha = 1 - (1 - \phi_D)^n \qquad (5\text{-}8)$$

$$\mu = \frac{1 - 2\left(\frac{Vs}{Vc}\right)^2}{2\left(1 - \frac{Vs}{Vc}\right)^2} \qquad (5\text{-}11)$$

$$I_{sh} = \frac{\phi_S - \phi_D}{\phi_S} \qquad (5\text{-}12)$$

$$\mu = A * I_{sh} + B \qquad (5\text{-}13)$$

Zero Tensile Strength Method, 5.17

$$\sigma_1' = S - P\text{, psi} \qquad (5\text{-}19)$$

$$F = \sigma_t + \sigma_1'\left(\frac{\mu}{1-\mu}\right) + P\text{, psi} \qquad (5\text{-}25)$$

$$\sigma_t = \sigma_1' * \beta\text{, psi} \qquad (5\text{-}28)$$

$$\frac{F}{D} = \left(\frac{S}{D} - \frac{P}{D}\right)\left[\left(\frac{\mu}{1-\mu}\right) + \beta\right] + \frac{P}{D}\text{, psi/ft} \qquad (5\text{-}29)$$

Kick Control, 5.34

$$\text{Reservoir pressure} = \text{SIDP} + W\ (\text{lb/gal}) * \text{depth (ft)} * 0.0519\ \text{psi} \tag{5-31}$$

$$\text{Kill mud density, lb/gal} = \frac{\text{reservoir pressure, psi}}{\text{depth, ft} * 0.0519} \tag{5-32}$$

$$\text{Initial circulating pressure} = \text{System pressure loss at kill rate} + \text{SIDP, psi} \tag{5-33}$$

$$\text{System pressure loss} = \text{Initial circulating pressure} - \text{SIDP, psi} \tag{5-34}$$

$$\text{Final circulating pressure} = \text{system pressure loss} * \frac{\text{kill mud density}}{\text{old mud density}}\text{, psi} \tag{5-35}$$

$$\text{Length of kick} = \frac{1029}{(\text{hole diameter}^2 - \text{collar O.D.}^2)} * \text{kick volume, bbl} \tag{5-36}$$

$$\text{Length of kick} = \text{length of collars} + \frac{1029}{(\text{hole diameter}^2 - \text{pipe O.D.}^2)} * (\text{kick vol} - \text{annular vol around collars})\text{, ft} \tag{5-37}$$

$$\text{SICP} - \text{SIDP} = \text{height of influx (ft)} * (\text{mud density, psi/ft} - \text{influx density psi/ft})\text{, psi} \tag{5-38}$$

$$\text{SICP} = \text{height of influx (ft)} * (\text{mud density, psi/ft} - \text{influx density, psi/ft}) + \text{SIDP, psi} \tag{5-39}$$

Density of influx = mud density (5-40)

$$- \frac{\text{SICP - SIDP}}{\text{height of influx}}, \text{ psi/ft}$$

Sacks of barite/100 bbl mud = 1490 (5-41)

$$* \frac{\text{(kill density - old density)}}{\text{3.58 - kill density}}$$

$$V = \frac{100\ (W_2 - W_1)}{35.8 - W_2} \tag{5-42}$$

Kick Tolerance, 5.39

$$SICP_{max} = 0.0519 * (FG_{min} - W) * D_F, \text{ psi} \tag{5-43}$$

$$K = \left[ \frac{D_F}{D_B} (FG_{min} - W) \right] + W, \text{ lb/gal} \tag{5-45}$$

$$K_{min} = \left[ \frac{D_F}{D_B} (FG_{min} - W) \right] - \frac{L_K}{D_B} (W - W_K) + W, \text{ lb/gal} \tag{5-46}$$

$$\Delta K = \frac{D_F}{D_B} (FG_{min} - W), \text{ lb/gal} \tag{5-47}$$

$$\Delta K_{min} = \left[ \frac{D_F}{D_B} (FG_{min} - W) \right] - \left[ \frac{L_K}{D_B} (W - W_K) \right], \text{ lb/gal} \tag{5-48}$$

# APPENDIX B
# Rw DETERMINATION FROM SP

By calculating $R_w$ for various water-bearing permeable zones, it is possible to accurately define the density of these pore waters by the use of simple conversion charts. This data is thus invaluable for providing precise normal hydrostatic gradients.

In brief, the procedure for determining fluid density from electric log data is as follows:

1) Determing the diameter of invasion from the appropriate $R_{int}$ chart (Schlumberger, "Log Interpretation Charts"). It will frequently be a sufficient approximation to assume no invasion, however.

2) Correct the SP reading for bed thickness and invasion, using chart B-1 or B-2. The results may be considered reliable only if the correction does not exceed 20%.

3) Convert Rmf at 75° F to Rmf at formation temperature, using nomogram B-3.

4) Convert Rmf to $Rmf_{eq}$. If Rmf at 75° F exceeds 0.1 ohm-m, use $Rmf_{eq}$=0.85 Rmf (at formation temperature). Otherwise, use chart B-4 for the conversion.

5) Find $Rw_{eq}$, using chart B-5.

6) Convert $Rw_{eq}$ to Rw, using chart B-4.

7) Convert Rw to salinity, using nonomgram B-3.

8) Deterine fluid density, using chart C-1.

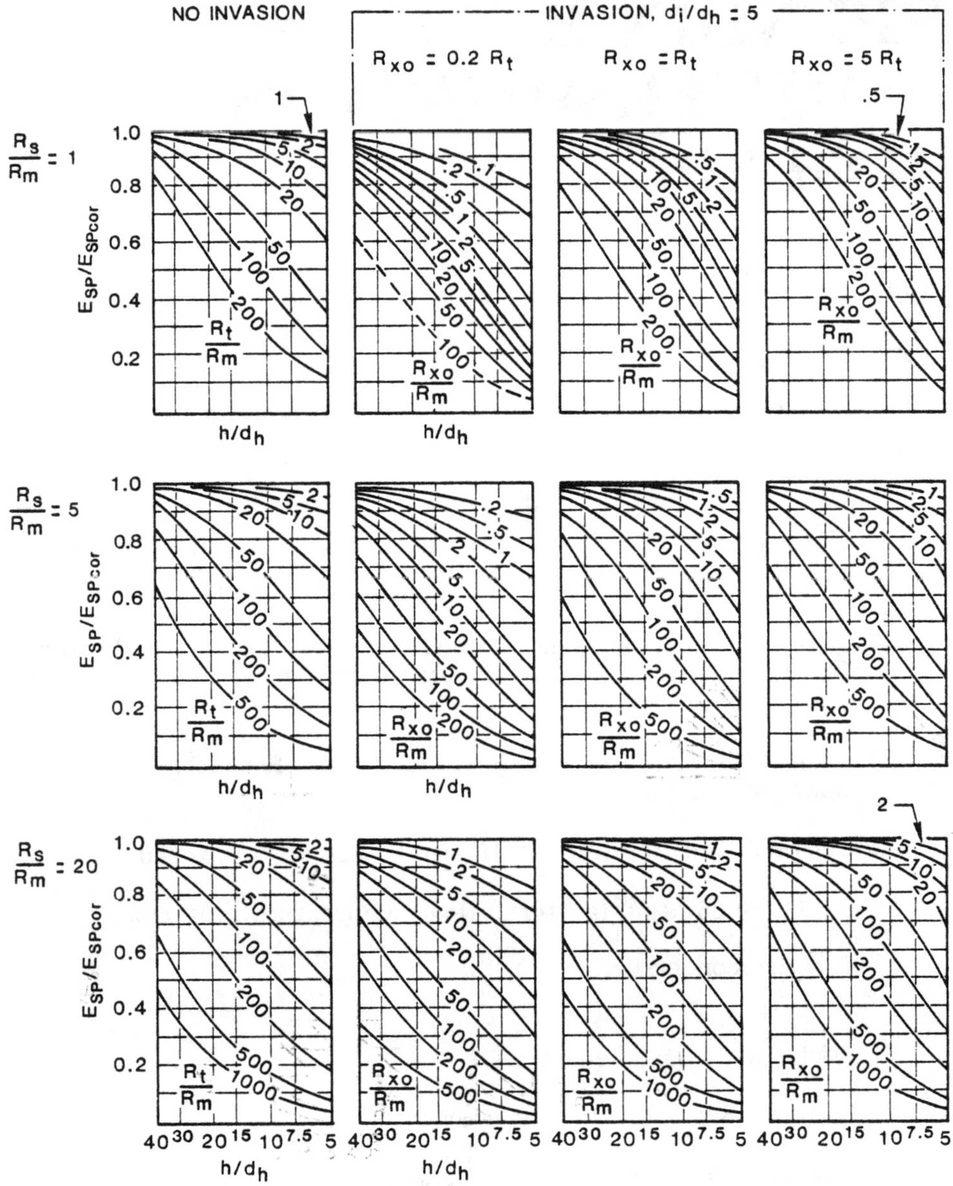


1. SELECT ROW OF CHART FOR MOST APPROPRIATE VALUE OF $R_s/R_m$.
2. SELECT CHART FOR NO INVASION OR FOR INVASION OF $d_i/d_h = 5$, AS MORE APPROPRIATE.
3. ENTER ABSCISSA WITH VALUE OF $h/d_h$ (RATIO OF BED THICKNESS TO HOLE DIAMETER).
4. GO VERTICALLY UP TO CURVE FOR APPROPRIATE $R_t/R_m$ (FOR NO INVASION) OR $R_{xo}/R_m$ (FOR INVADED CASES), INTERPOLATING BETWEEN CURVES IF NECESSARY.
5. READ $E_{SP}/E_{SPcor}$ IN ORDINATE SCALE. CALCULATE $E_{SPcor} = E_{SP}/(E_{SPcor})$. $E_{SP}$ IS SP FROM LOG.

Figure B-1. SP Correction Charts (for representative cases)

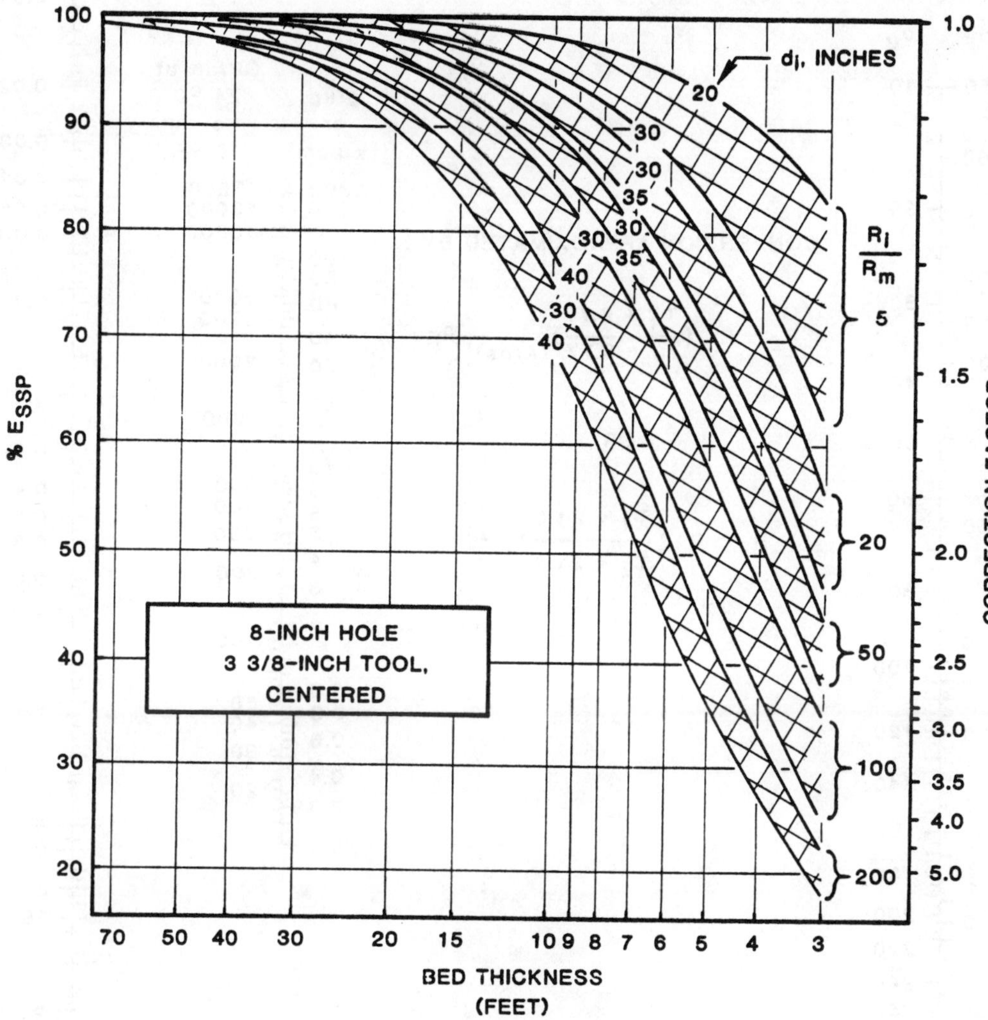


Figure B-2. SP Correction Charts (empirical)

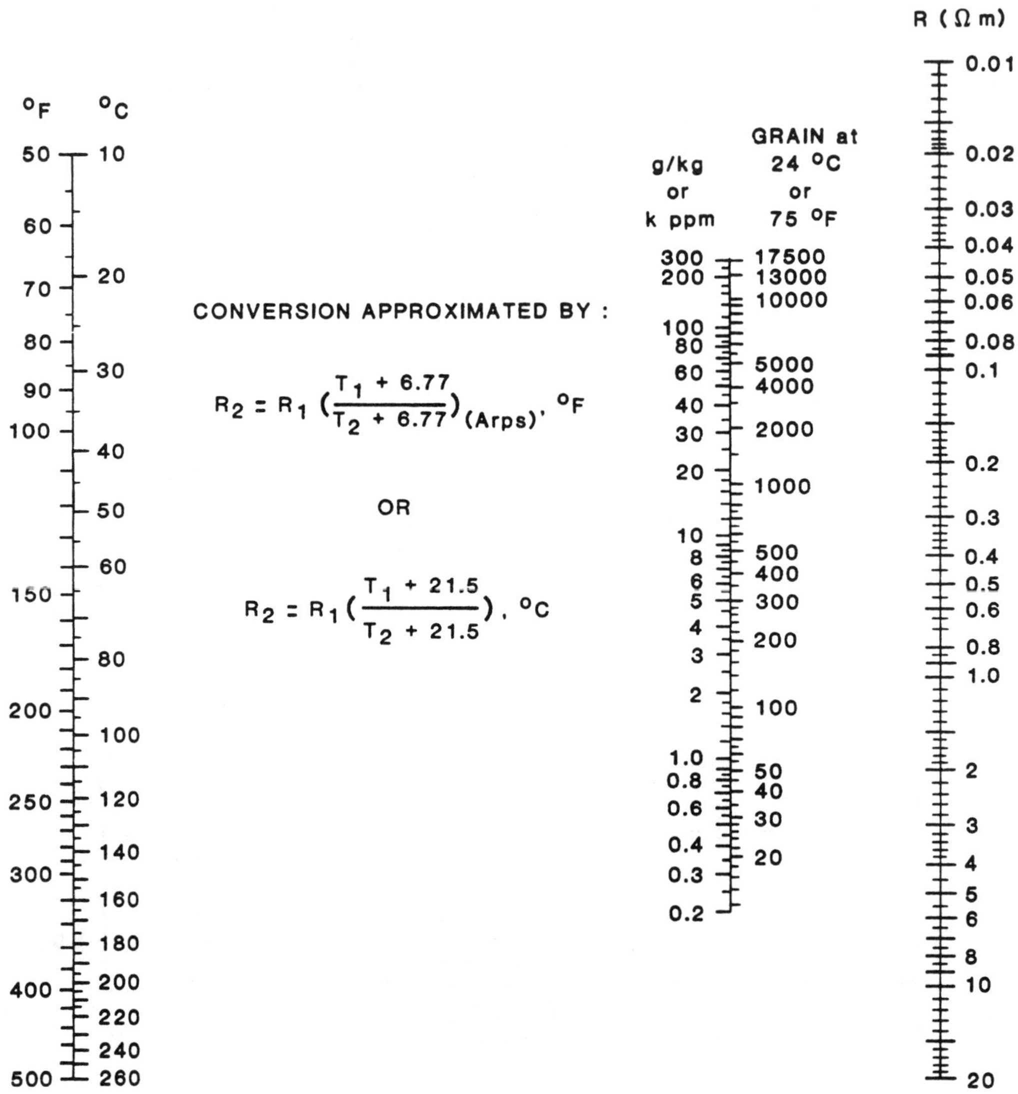


Figure B-3. Resistivity nomograph for NaCl solutions

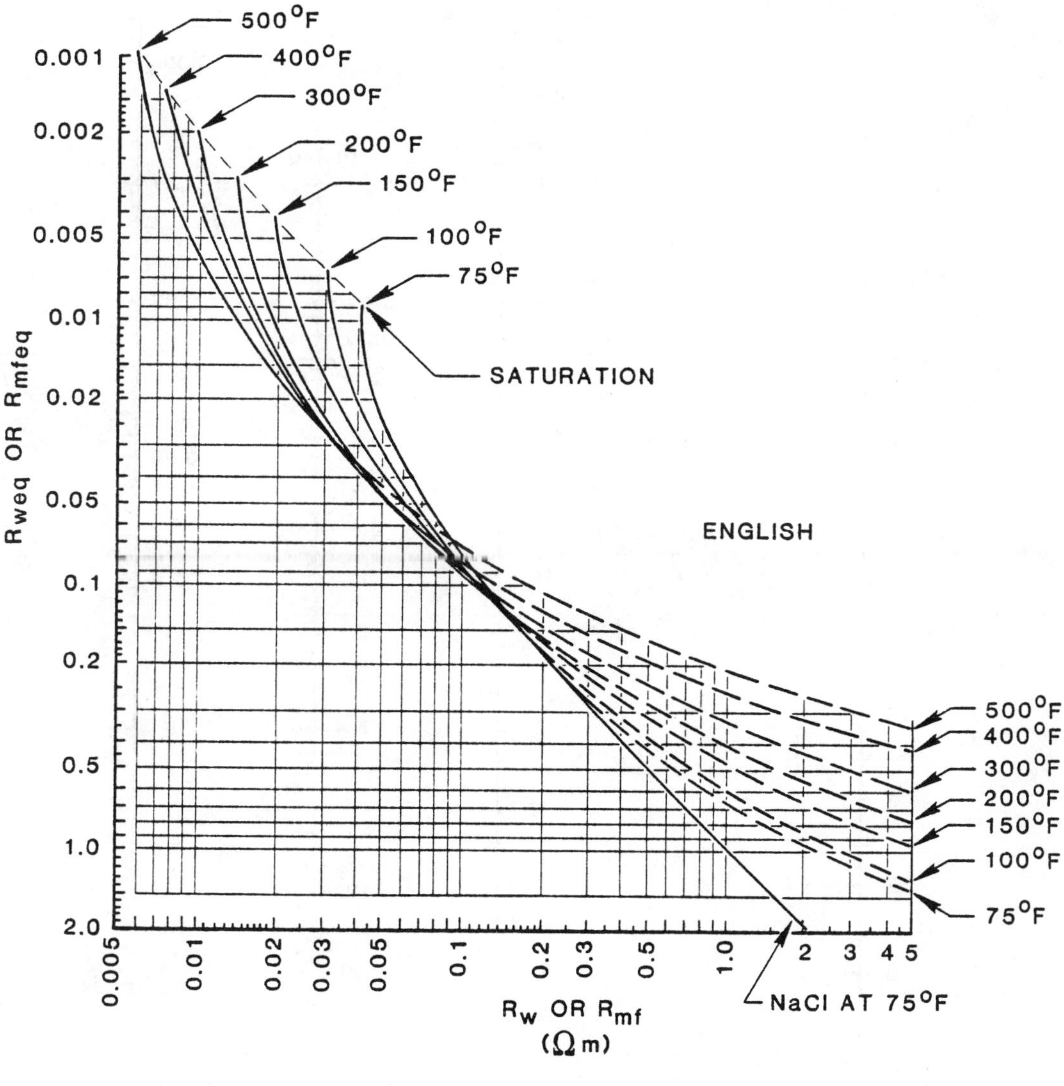


Figure B-4. Rw versus $Rw_{eq}$ and formation temperature

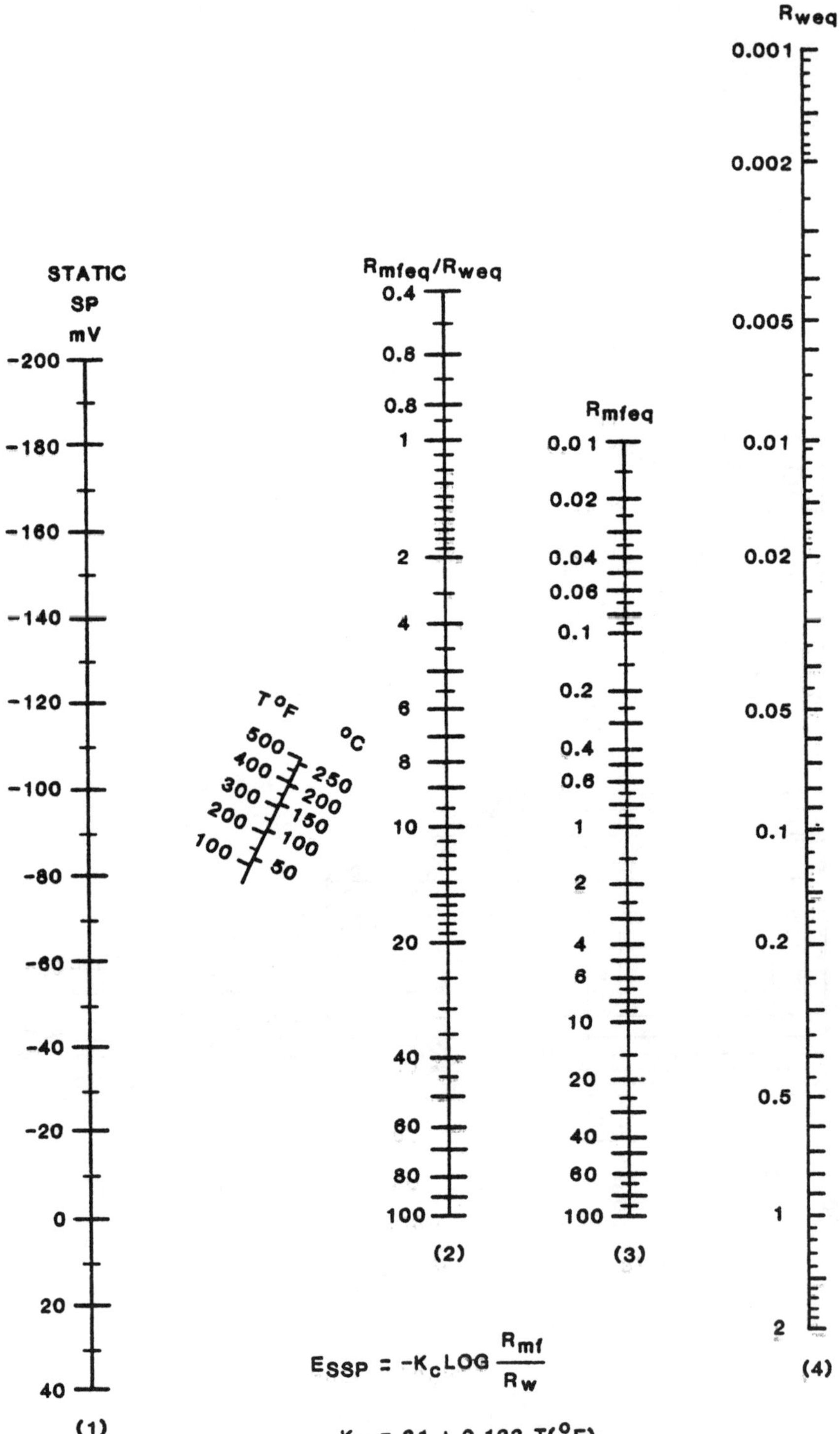


$$E_{SSP} = -K_c \text{LOG} \frac{R_{mf}}{R_w}$$

$$K_c = 61 + 0.133\ T(^{\circ}F)$$

$$K_c = 65 + 0.24\ T(^{\circ}C)$$

Figure B-5. $Rw_{eq}$ determination from the SSP (clean formation)

# APPENDIX C
# NOMOGRAMS

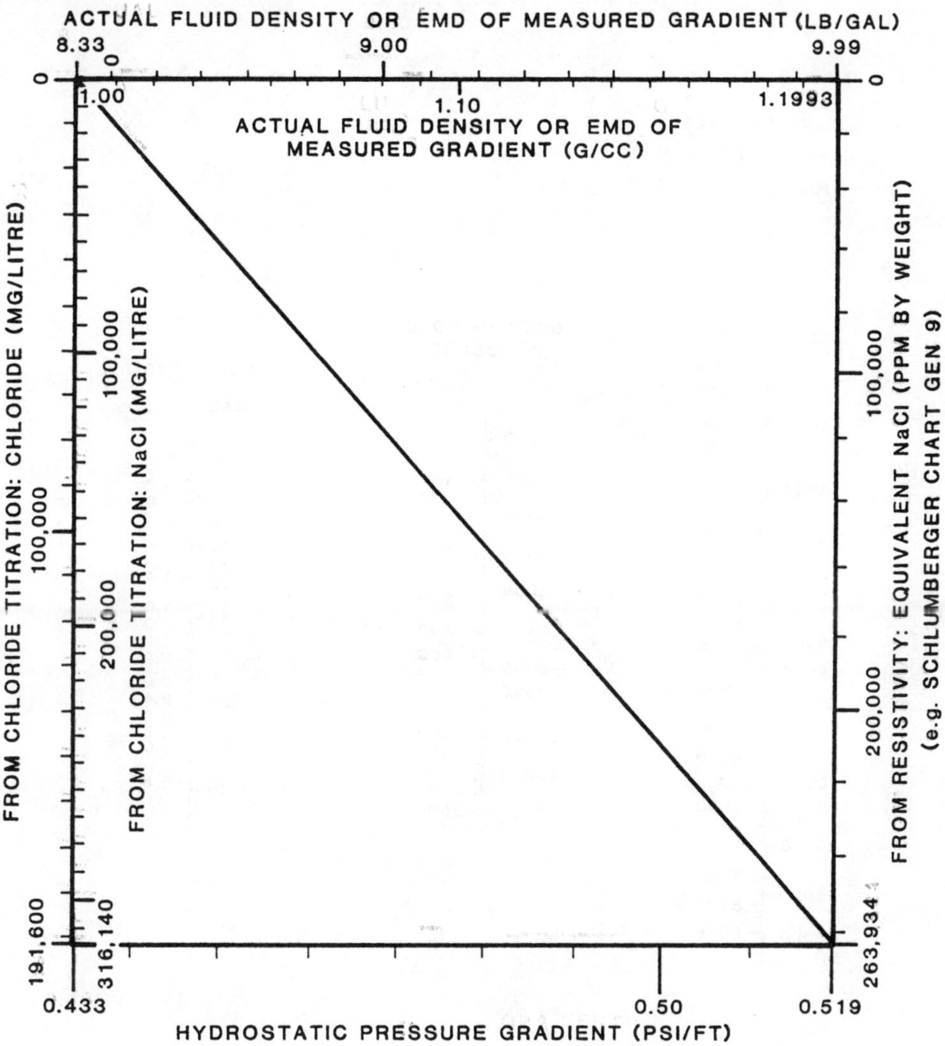


NaCl (PPM M BY VOL)..CHLORIDE (MG/LITRE) X 1.65

NaCl (PPM BY WT)......$\dfrac{\text{CHLORIDE (MG/LITRE) X 1.65}}{\text{ACTUAL FLUID DENSITY (g/cc)}}$

DENSITY (LB/GAL).............. DENSITY (g/cc) X 8.34

EMD (LB/GAL) ....MEASURED GRADIENT (PSI/FT) 0.0519

Figure C-1. Formation fluid salinity and density

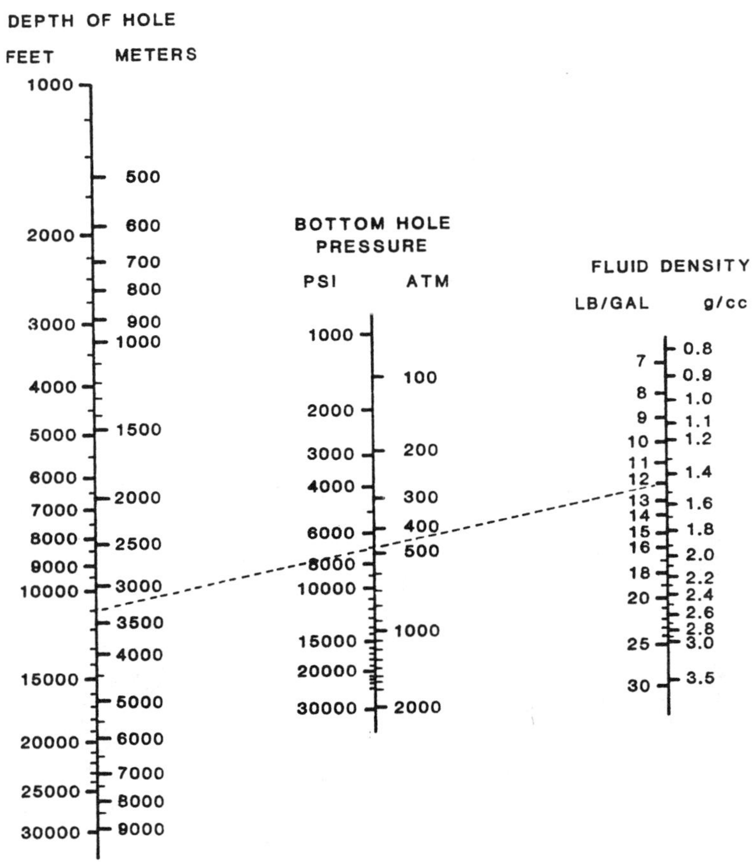


Figure C-2. Hydrostatic pressure of fluid columns

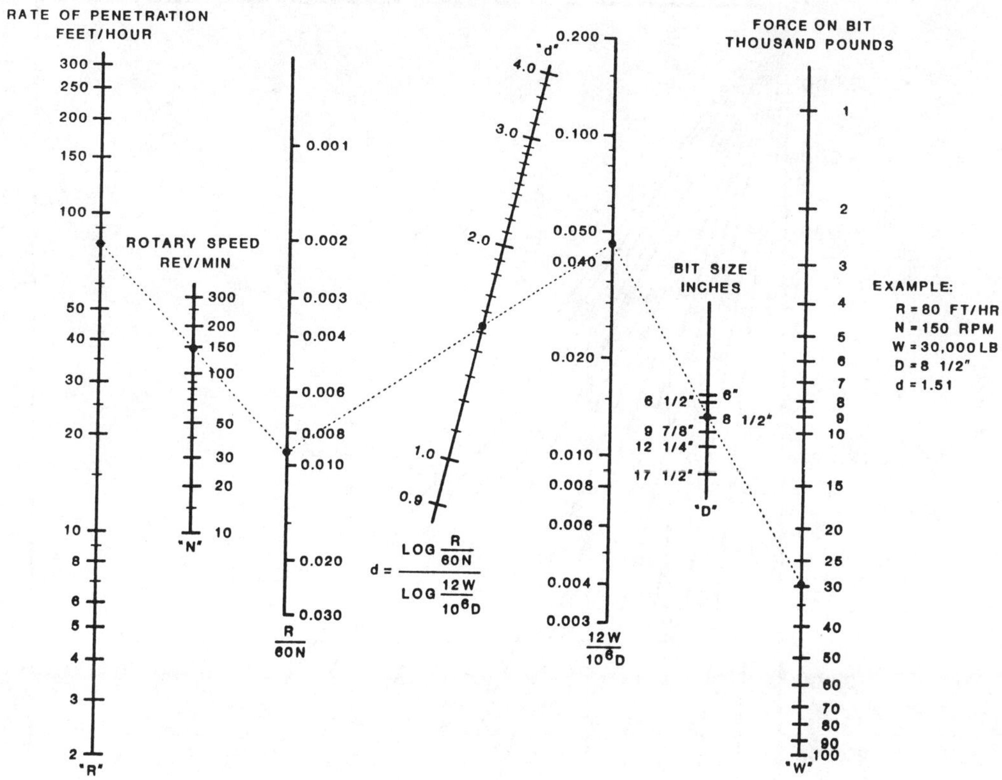


Figure C-3. $D_{xc}$ nomogram

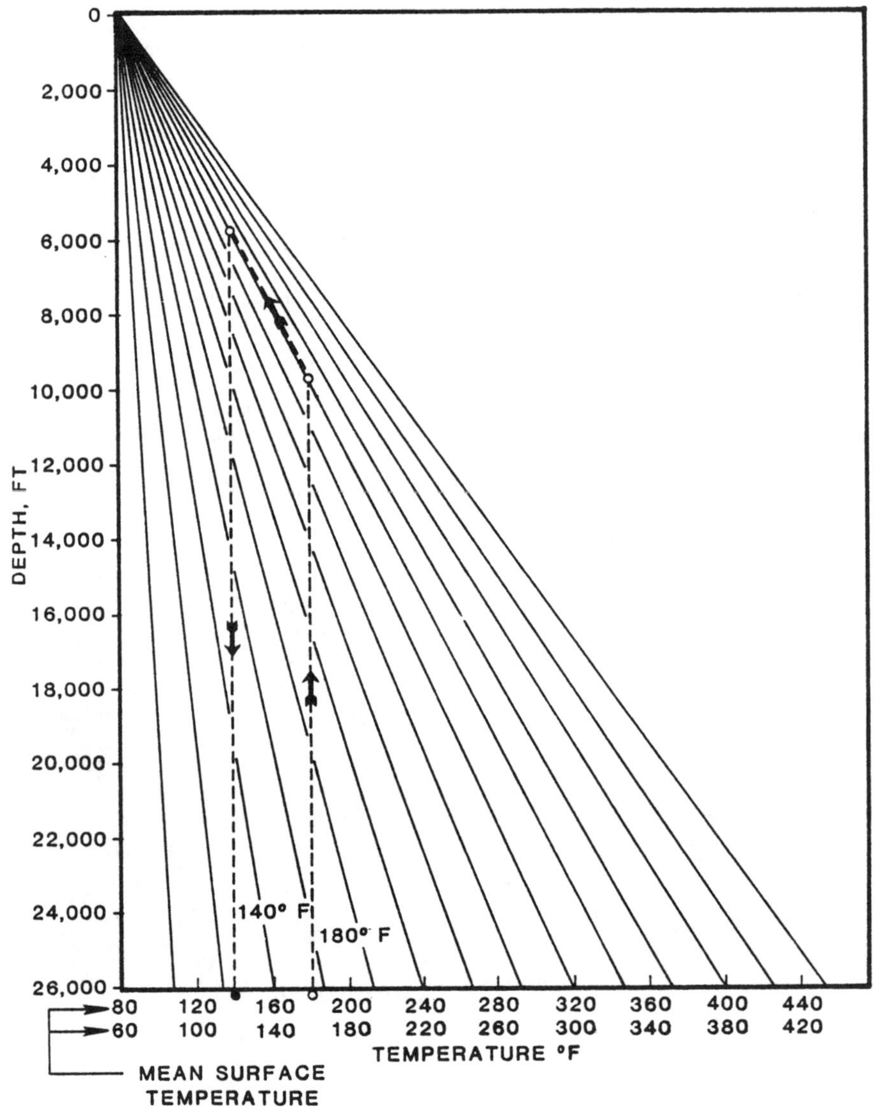


Figure C-4. Geothermal gradient and bottomhole temperature

# GLOSSARY

**Abnormal pore pressure:** Pressure contained within a pore system that is in excess of the normal hydrostatic pore pressure. General usage is limited to description of excess pressure (see Subnormal pore pressure).

**Allochthon:** A mass of rocks that has been moved from its site of origin by tectonic forces, as in a thrust sheet or nappe; i.e., of foreign origin, or introduced.

**Annular pressure loss:** That pressure which is necessary to overcome the frictional forces between the annulus, drilling assembly and drilling fluid.

**Aquathermal pressure:** A term proposed by Barker (1972) describing a hypothetical geopressure mechanism. If pore volume remains constant with burial and temperature increase, the thermal expansion of water (being approximately 300 times that of typical sedimentary minerals) can cause extremely rapid pore pressure increase. Water density, by definition, must remain constant.

**Aquifer:** A body of rock that contains sufficient saturated permeable material to conduct groundwater and to yield significant quantities of groundwater to wells and springs.

**Blowout:** Loss of control of a well due to an uncontrolled kick.

**Cap rock:** Originally defined to describe that rock overlying the top of a salt body, composed of anhydrite and gypsum, with minor calcite and sulfur, resulting from accumulation of the less soluble minerals of the salt body during leaching of its top. The term was used in pressure evaluation to provide an explanation for entrapment of pore waters during burial, providing a seal, thus allowing pore pressure to increase. It has been found that cap rocks are the exception rather than the rule: examination of the principles involved will show that, for a cap rock to form, a geopressure must have existed previously and undergone leakage in order to precipitate minerals above the anomaly.

**Cation exchange capacity (C.E.C.):** A measure of the total amount of exchange cations of a mineral. Exchange sites are most prolific in clay minerals, particularly the smectite group. Actual cation exchange capacity varies with particle size and with the nature of the cation. Kaolinites possess a higher anion exchange capacity than cation exchange capacity.

**Compaction disequilibrium:** Synonymous with subcompaction is a process by which the delicate balance between rate of sedimentation, burial, porosity reductions and expulsion of pore fluids become upset by a change in any of the contributing factors, resulting in a pore pressure increase. Overall, a pressure increase is caused by the effective decrease in dewatering efficiency.

**Degraded illite:** Illite that has had much of its potassium removed from the interlayer position as a result of leaching.

**Diapirism:** The process of rupturing domed or uplifted rocks by plastic core material, caused by the effect of geostatic load or large density differences.

**Differential pressure:** At any point in the wellbore, whether the mud is circulating or static, it is the difference between the pore pressure and the pressure exerted by the mud column. Overbalance occurs when the mud pressure is greater than the pore pressure, and underbalance occurs when the mud pressure is less than the pore pressure.

**Drillability:** Describes the interaction between a particular bit in a particular lithology. Thus, when the correct bit is used for a particular lithology, rate of penetration is proportional to drillability.

**Effective circulating density (ECD):** ECD is the combination of the hydrostatic pressure of the mud in a static condition, plus the frictional forces caused by mud moving up the annulus: the annular pressure loss. Converting the sum of the pressures to a gradient gives the total effective mud density (EMD) at TD.

**Effective overburden Pressure ($\sigma_1'$):** The difference between the total overburden pressure and the pore pressure at any particular point in the formation is the effective overburden pressure. It is a stress that acts vertically downwards. It is this stress that is largely responsible for compaction. Effective overburden pressure is also referred to as matrix stress, rock-grain stress, and rock-skeleton stress.

**Effective permeability:** The observed permeability of a porous medium to one fluid phase which is under conditions of physical interaction (i.e., friction, surface tension) with another fluid phase present in the same pore system.

**Effective stress ($\sigma'$):** Any principle stress, tensional or compressive, minus the pore pressure.

**Elastic:** Describes the ability of a material to return to its original shape and dimensions when the deforming forces are removed.

**Electro-osmosis:** The motion of a liquid through a membrane under the influence of an applied electric field.

**Equivalent Mud Density (EMD):** A convenient reference by which any downhole or subsurface pressure, when converted to a gradient referenced to the flowline, describes the equivalent mud density that would produce that particular pressure at that particular depth.

**Failure envelope:** An envelope of a series of Mohr circles; the locus of points whose coordinates on a differential stress/shear stress plot represent the stresses causing failure. Failure envelope is synonymous with Mohr envelope and rupture envelope.

**Finite Strain:** The total amount of strain (deformation) recorded by a particular structure, irrespective of episodic deformational events.

**Formation balance gradient (FBG):** The formation pore pressure gradient at a particular point referenced to the flowline. Because of the air gap and water depth, the FBG offshore is always less than the actual pore pressure gradient, becoming asymptotic at depth.

**Formation-volume factor:** The volume of a liquid at reservoir conditions divided by the volume at surface conditions.

**Fracture pressure:** Is the pressure in the borehole at which whole mud is injected into the formation due to the initiation and extension of natural and pressure-induced fractures.

**Geopressure:** A term introduced by Stuart, describing any porous formation in which the pore pressure is in excess of the normal hydrostatic pressure (see Abnormal pore pressure).

**Hydraulic conductivity:** The rate of flow of water through unit cross-sectional area under unit hydraulic gradient at the prevailing temperature. Synonymous with permeability coefficient.

**Hydrostatic pressure:** The pressure exerted by the water (fluid) at any given point in a body of water (fluid) at rest. The hydrostatic pressure of groundwater is generally due to the density of the water and the vertical height of the water column.

**Illite:** A general name for a group of three-layer, mica-like clay minerals which are intermediate in composition and structure between muscovite and montmorillonite, and which have 10-angstrom c-axis spacings that show essentially no lattice expanding characteristics. Illite contains less potassium and more water than true micas.

**Interval transit time:** The reciprocal of sonic compressional wave velocity over a fixed distance, measured in micro-seconds per foot.

**Ionic filtration:** A process of concentrating ions on one side of a semipermeable membrane as fluid passes through the membrane. The efficiency of the membrane in restricting ions or certain ions is a function of clay mineralogy, pore geometry, porosity, etc.

**Isotropic:** Describes a medium, the properties of which are the same in all directions.

**Kick:** An unexpected influx of formation fluids into the borehole that may be controlled by closing the blowout preventers.

**Kick tolerance:** Estimated as the maximum pressure or the mud density that the weakest part of the borehole (formation, casing or surface equipment) can withstand, in the event that a kick is taken.

**Laminar flow:** Fluid flow in which the streamlines remain distinct and in which the flow direction at every point remains unchanged with time.

**Matrix stress:** Synonymous with effective overburden stress or pressure.

**Metamontmorillonite:** A term used to describe clay minerals which are composed of montmorillonite, muscovite and illite in varying proportions within the same aggregate; i.e., montmorillonite interlayered with muscovite and illite.

**Montmorillonite:** A member of the smectite group of swelling clays. Montmorillonite is a subgroup of expanding lattice clay minerals characterized by a three-layer crystal lattice, by deficiencies in charge in the tetrahedral and octahedral positions balanced by cations subject to exchange, and by swelling or wetting due to the adsorption of considerable interlayer water. Montmorillonites are the chief constituents of bentonite. In some terminology, montmorillonite and smectite are synonymous.

**Montmorillonite dehydration:** A hypothetical geopressure-generating mechanism (initiated by temperature) that involves the release of structured monomolecular hydrogen-bonded water from montmorillonite interlayer sites to the pores, resulting in a net expansion of the water as it undergoes the phase change. Experimental evidence has shown that structured water has slightly higher density than normal water so that, upon desorption, the released water expands — resulting in pressure increase in the closed pore system.

**Mylonitization:** Deformation of a rock by extreme micro-brecciation without chemical reconstitution of the granulated minerals. Characteristic appearance is flinty, banded or streaked, and may contain undestroyed augen of the parent rock in a granulated matrix.

**Normal formation balance gradient (N.FBG):** The normal pore pressure gradient referenced to the flowline.

**Normal pore pressure:** Synonymous with hydrostatic pressure when referring to pore waters only.

**Osmosis:** The spontaneous movement of water through a semipermeable membrane which separates two solutions of different concentrations, until the concentration of each solution becomes equal.

**Overburden pressure (S):** The total vertical stress exerted by the weight of the overlying rocks and their contained fluids.

**Permeability:** A measure of the relative ease of fluid flow under unequal pressure; normal unit of measurement is the millidarcy (md).

**Piezometric (potentiometric) surface:** An imaginary surface representing the static head of groundwater and defined by the level to which water will rise in a well. The water table is a particular potentiometric surface.

**Poisson's ratio ($\mu$ or $\upsilon$):** The ratio of the lateral unit strain to the longitudinal strain in a body that has been stressed longitudinally within its elastic limit.

**Pressure potential:** In an aquifer, the rate of change of pressure head per unit of distance of flow at a given point and in a given direction. Synonymous with hydraulic gradient, hydraulic potential.

**Pseudotachylite:** A dense rock produced in the compression and shear conditions associated with intense and extensive fault movements, involving extreme mylonitization and partial melting. Frictional melting occurs when water is absent, and the expansion upon the phase change allows the resultant glass to be intrusive.

**Semipermeable membrane:** A membrane that is partially but not freely or wholly permeable to particular solutions.

**Smectite:** A clay group containing the minerals montmorillonite, beidellite, nontronite, saponite, hectorite and sauconite. All are swelling clay minerals.

**Subcompaction:** See compaction disequilibrium.

**Subnormal pore pressure:** That pressure contained within a pore system that is less than normal hydrostatic pore pressure.

**Tectonic stress:** An additional applied stress, independent of gravity stresses, that is responsible ultimately for producing tectonic deformation; structures.

**Undercompacted:** Compaction of sedimentary rock less than that normal for the existing overburden pressure. Synonomous with underconsolidation. Refer to compaction disequilibrium.

**Weight:** The force produced by the action of gravity on a mass.

# REFERENCES

Anderson, E. M., 1942, The Dynamics of Faulting, Oliver and Boyd, London.

Anderson, R. A., D. S. Ingram and A. M. Zanier, 1973, Determining Fracture Pressure Gradients from Well Logs, J. P. T., Nov.

Barker, C., 1972, Aquathermal Pressure — Role of Temperature in Development of Abnormal-Pressure Zones, AAPG Bull., v. 56, n. 10.

Bellotti, P., and D. Giacca, 1978, Pressure Evaluation Improves Drilling Programs OG& J., Sept. 11

Bingham, M. G., 1965, A New Approach to Interpreting Rock Drillability, The Petroleum Publishing Co.

Bradley, J. S., 1975, Abnormal Formation Pressure, AAPG Bull., v. 59, n. 6.

Bradley, W. B., 1979 (a), Mathematical Concept -- Stress Cloud -- Can Predict Borehole Failure, O&G J., Feb.

Bradley, W. B., 1979, Predicting Borehole Failure Near Salt Domes, O&G J., Apr.

Brederhoeft, J. D. et al, 1963, Possible Mechanisms for Concentration of Brines in Subsurface Formations, AAPG Bull., v. 47, n. 2.

Burst, J. F., 1969, Diagenesis of Gulf Coast Clayey Sediments and Its Possible Relation to Petroleum Migration, AAPG Bull., v. 53, n. 1.

Chenevert, M. E., and L. J. McClure, 1978, How to Run Casing and Open Hole Pressure Tests, O&G J., Mar.

Christman, S. A., 1973, Offshore Fracture Gradients, J. P. T., SPE Paper 4133.

Costley, R. D., 1967, Hazards and Costs Cut by Planned Drilling Programs, World Oil, Oct.

Deer, W. A. et al, 1967, An Introduction to the Rock Forming Minerals, Longmans Press.

Dowdle, W. L., and W. M. Cobb, 1975, Static Formation Temperature from Well Logs, J. P. T., Nov.

Eaton, B. A., 1969, Fracture Gradient Prediction and Its Application in Oilfield Operations, J. P. T., Oct.

Fairhurst, C., 1960, On the Determination of the State of Stress in Rock Masses, SPE Paper 1062.

Fertl, W. H., 1973, Abnormal Formation Pressures, Elsevier Press.

Fowler, P. T., 1980, Telling Live Basins from Dead Ones By Temperature, World Oil, May.

Fyfe, W. S., N. J. Price and A. B. Thompson, 1978, Fluids in the Earth's Crust, Elsevier Scientific Pub. Co.

Goins, W. D. Jr., 1968, Guidelines for Blowout Prevention, World Oil, Oct.

Goldsmith, R. G., 1972, Why Gas-Cut Mud is Not Always a Serious Problem, World Oil, v. 175, no. 5.

Gretener, P. E., 1978, Pore Pressure: Fundamentals, General Ramifications and Implications for Structural Geology, AAPG Course Note Series 4.

Hafner, W., 1951, Stress Distributions and Faulting, Bull. Geol. Soc. Am., v. 62.

Haimson, B., and C. Fairhurst, 1969, Hydraulic Fracturing in Porous-Permeable Materials, J. P. T., SPE Paper 2354.

Haimson, B., and J. N. Edl, 1972, Hydraulic Fracturing of Deep Wells, SPE Paper 4061.

Handin, J. et al, 1963, Experimental Deformation of Sedimentary Rocks Under Confining Pressure: Pore Pressure Tests, AAPG Bull., v. 47, n. 5.

Hanshaw, B. B., and E. Zen, 1965, Osmotic Equilibrium and Overthrust Faulting, Geol. Soc. Am. Bull., v. 76, n. 12.

Hinch, H. H., 1978, The Nature of Shales and the Dynamics of Hydrocarbon Expulsion in the Gulf Coast Tertiary Section, AAPG Course Note Series 8.

Hottman, C. E., and R. K. Johnson, 1965, Estimation of Formation Pressures from Log-Derived Shale Properties, J. P. T., Jun.

Hubbert, M. K., 1945, Strength of the Earth, AAPG Bull., v. 29, n. 11.

Hubbert, M. K., 1972, Structural Geology, Hafner Pub. Co.

Hubbert, M. K., and D. G. Willis, 1957, Mechanics of Hydraulic Fracturing, Trans. AIME, v. 210.

Jaeger, J. C., and N. G. W. Cook, 1976, Fundamentals of Rock Mechanics, Chapman and Hall, 2nd ed., London.

Jeffreys, H., 1952, The Earth, Cambridge, 3rd ed.

Jones, P. H., 1969, Hydrodynamics of Geopressures in the Northern Gulf of Mexico Basin, J. P. T., July.

Jorden, J. R., and O. J. Shirley, 1966, Application of Drilling Performance Data to Overpressure Detection, J. P. T., Nov.

Kendall, H. A., 1977, How to Control Deep Critical Wells, Pet. Eng., Mar.

Kennedy, G. C., and W. T. Holser, 1966, Pressure-Volume-Temperature and Phase Relations of Water and Carbon Dioxide, Geol. Soc. Am. Memoir 97.

Lewis, C. R., and S. C. Rose, 1970, A Theory Relating High Temperatures and Overpressures, J. P. T., January.

MacPherson, L. A., and L. N. Berry, 1972, Prediction of Fracture Gradients from Log Derived Moduli, The Log Analyst, Sep-Oct.

Magara, K., 1975(b), Reevaluation of Montmorillonite Dehydration as Cause of Abnormal Pressure and Hydrocarbon Migration, AAPG Bull., v. 59, n. 2.

Magara, K., 1975(a), Importance of Aquathermal Pressuring Effect in Gulf Coast, AAPG Bull., v. 59, n. 10.

McCain, W. D., 1973, The Properties of Petroleum Fluids, The Petroleum Publishing Company.

Matthews, W. R., and J. Kelly, 1967, How to Predict Formation Pressure and Fracture Gradient, O&G J., Feb. 20

Nwachukwu, S. O., 1976, Approximate Geothermal Gradients in Niger Delta Sedimentary Basin, AAPG Bull., v. 60, n. 7.

Olsen, H. W., 1972, Liquid Movement Through Kaolinite under Hydraulic, Electric and Osmotic Gradients, AAPG Bull., v. 56, n. 10.

Pennebaker, E. S., 1968, An Engineering Interpretation of Seismic Data, SPE Paper 2165.

Pilkington, P. E., and H. A. Neihaus, 1975, Exploding the Myths About Kick Tolerance, World Oil, Jun.

Pilkington, P. E., 1978, Fracture Gradient Estimates in Tertiary Basins, Pet. Eng., May.

Powers, M. C., 1967, Fluid-Release Mechanisms in Compacting Marine Mudrocks and their Importance in Oil Exploration, AAPG Bull., v. 51, n. 7.

Rehm, B., 1976, Deep Water Drilling Poses Special Pressure Control Problems, O&G J, May 3.

Rehm, B., and R. McClendon, 1971, Measurement of Formation Pressure from Drilling Data, SPE 3601, SPE Reprint Series No. 6a, 1973 revision.

Reynolds, E. B., 1970, Predicting Overpressured Zones with Seismic Data, World Oil, v. 171, Oct.

Rieke III, H. H., and G. V. Chilingarian, 1974, Compaction of Argillaceous Sediments, Developments in Sedimentology, 16, Elsevier Press, New York.

Rubey, W. W., and M. K. Hubbert, 1959, Overthrust Belt in Geosynclinal Area of Western Wyoming in Light of Fluid Pressure Hypothesis, Geol. Soc. Am. Bull., v. 70.

Taylor, D. B., and T. K. Smith, 1970, New Fracture Gradients Help Cut Costs Offshore, World Oil, Jun.

U. S. Bureau of Reclamation, 1953, Physical Properties of Some Typical Foundation Rocks, Concrete Laboratory Rpt. SP-39.

Weaver, C. E., and K. C. Beck, 1971, Clay Water Diagenesis During Burial: How Mud Becomes Gneiss, Geol. Soc. Am. Special Paper 134.

Weurker, R. G., 1963, Annotated Tables of Strength and Elastic Properties of Rocks, SPE Reprint Series, n. 6.

Young, A., and P. F. Low, 1965, Osmosis in Argillaceous Rocks, AAPG Bull., v. 49, n. 7.

# INDEX